BestMasters

Springer awards „BestMasters" to the best master's theses which have been completed at renowned universities in Germany, Austria, and Switzerland.

The studies received highest marks and were recommended for publication by supervisors. They address current issues from various fields of research in natural sciences, psychology, technology, and economics.

The series addresses practitioners as well as scientists and, in particular, offers guidance for early stage researchers.

Markus Merz

Scarce Natural Resources, Recycling, Innovation and Growth

 Springer Gabler

Markus Merz
Tübingen, Germany

BestMasters
ISBN 978-3-658-12054-2   ISBN 978-3-658-12055-9 (eBook)
DOI 10.1007/978-3-658-12055-9

Library of Congress Control Number: 2015955646

Springer Gabler

Printed on acid-free paper

Springer Gabler is a brand of Springer Fachmedien Wiesbaden
Springer Fachmedien Wiesbaden is part of Springer Science+Business Media
(www.springer.com)

Contents

Appendix: A Sketch of Solutions

List of Figures

List of Symbols

A	Technological progress parameter
C	Consumption
c	Consumption in efficiency units
$\tilde{c}$	Consumption in efficiency units adjusted by the resource input
D	Stock of waste
$F(\cdot)$	Production technology
F_K	Marginal product of capital
F_R	Marginal product of the exhaustible resource
g	Growth rate
H	Hamiltonian operator
$h(z)$	Conditional probability of completion
I	Investment
J	Flow of secondary material input (recyclable resource)
K	Capital
k	Capital in efficiency units
$\tilde{k}$	Capital in efficiency units adjusted by resource input
L	Labor
MRS	Marginal rate of substitution
m	Effort in $R\&D$
N	Constant extraction of the durable commodity
$P(\cdot)$	Alternative production technology
Q	Inventory of the durable commodity

R	Flow of the exhaustible resource extraction
S	Stock of the exhaustible resource
$U(C)$	Utility of consuming C
V	Ratio of resource utilization to the stock of the resource
$v(m)$	*R&D* cost function
W	Additional utility stream
Y	Output
y	Output in efficiency units
$\tilde{y}$	Output in efficiency units adjusted by the resource input
x	Capital to resource ratio
Z	Utilization rate of the durable commodity
z	Cumulative effort in *R&D*
α	Output elasticity of capital
β	Output elasticity of the exhaustible resource
γ	Inter-temporal elasticity of substitution
δ	Growth rate of the technological progress
θ	Parameter which governs the elasticity of inter-temporal substitution
κ	Consumption share of output (identical to the reflux rate in the considered model)
μ	Material intensity of output
ρ	Time preference rate
σ	Elasticity of substitution
υ	Discount rate independent of the path followed by the economy
ϕ	Output elasticity of the recycled resource
χ	Fractional loss of material through use
ψ	Costate variable
ω	Probability that at time t the technological breakthrough has been reached
Ω	Probability that at time t the technological breakthrough has not been reached

1 Introduction

Natural resources are vital for the production of output. Economic activity is dependent on resources, i.e. land to grow food, raw materials to make goods, and energy to power machines. For example, according to the World Resource Institute (2010) about 37% of the world's land is used for agriculture and according to BP (2014) about 91 million barrels (bbls) of oil are consumed every day.[1] Oil plays a vital role in many aspects of daily life. Its components are used to produce almost all chemical products, such as plastic, detergent, paint, and even medicine. Its most apparent use is as a fuel for cars and airplanes. Natural resources are not created through a deliberate investment but rather exist irrespective of human activity. The opposite is true for physical and human capital.

The sustainability of natural resources is distinguished between renewable and nonrenewable. A nonrenewable resource exists in a finite quantity on the earth and has no natural regeneration process within a relevant time scale. Examples are fossil fuel, rare earths, ore, and other metals. The supply of a renewable resource is limited, however its stock is replenished by a timely natural regeneration process. If the stock of a renewable resource is not over-exploited, then it can be sustained. Wood is an example.

Increasing population and the capacity limits of earth may result in the exhaustion of renewable resources. Exhaustion is imminent for nonrenewable resources. Once a nonrenewable resource is consumed it is gone forever. Since the industrial era the economy is reliant primarily on exhaustible resources, specifically oil.

---

[1] A barrel, the unit of oil measurement, is equal to ≈ 159 liters.

This thesis evaluates whether the limited availability of the nonrenewable resources constrains the economy's (world's) growth potential. The theoretical evaluation is done using neoclassical growth models with resource constraints. The effect of recycling and technological change on the resource constraints is considered.

The work begins with an introduction into economic growth theory and provides an overview of the oil market. In chapter 3 basic economic growth theory is introduced. The well-known Dasgupta-Heal model is presented and analyzed. To avoid the resulting starvation of mankind, recycling as an intermediate solution is introduced in chapter 4. Chapter 5 presents technological progress as a long-term solution to economic growth. Resource-augmenting and backstop technology are analyzed. After a thorough analysis of the models it is concluded that the ultimate solution to long-term economic growth is a backstop technology.

2 An Introduction to Economic Growth Theory and the Oil Market

The reliance of production and economic growth on natural resources, specifically oil, may result in a decline on economic growth due to resource constraints.

Malthus (1803) was one of the first who attempted to answer the question whether natural resources hinder or even limit economic growth. He concluded that an exponential increase in population cannot be sustained with a modest linear increase of food production. The finite agricultural production possibilities due to limited land availability would result in the starvation of the excess population. The scarce natural resource, land, leads to long-term economic stagnation. Shortly after Malthus described his theory the industrial revolution began. The industrialization helped western European countries to escape the Malthusian trap.

In the pre-industrial era, the economy was based primarily on land and renewable resources, e.g. agricultural crops, wood, water, and wind. This changed dramatically with the industrial revolution. The economy is now based more heavily on exhaustible resources, initially coal and now oil. The finite nature of nonrenewable resources implies that an economy faces a trade-off between the present and the future. What effect the exhaustion of resources has on economic growth and welfare became relevant.

Starting with Hotelling (1931) various economists studied the optimal extraction of exhaustible resources and the resulting limits to economic growth. According to the basic Hotelling rule, the extraction of nonrenewable resources is most socially and economically profitable when the price of the resource increases with the rate of interest (no arbitrage condition). Overall social welfare requires that the consumption of resources in production today is balanced with the consumption of future production. This would result in the partial extraction of the resource over time. It, however, has been empirically proven that there are many other factors affecting the price of natural resources.[2] A prominent example is crude oil. In the last century the real price of oil has fluctuated significantly.

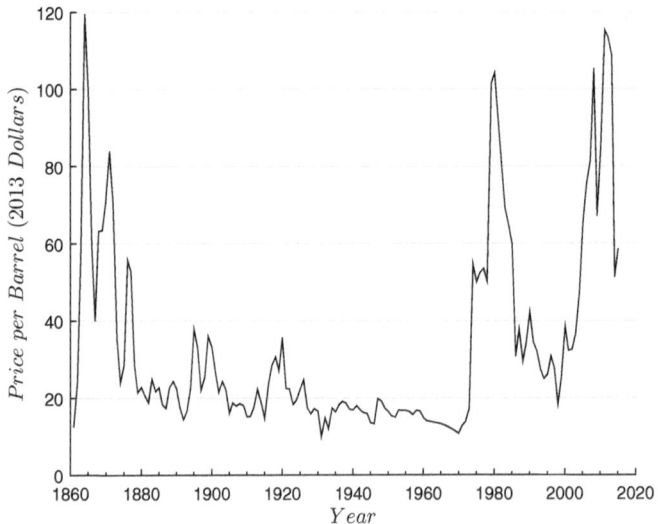

Fig. 2.1: Real Price of Oil, 1861-2015.
Source: BP (2014) Statistical Review of World Energy, extended to 2015.

The basic Hotelling's rule would suggest an increasing price behavior. The oil price, however, has been more dependent to political factors and to the market power of OPEC (Organization of the Petroleum Exporting Countries, the intergovernmental cartel that controls about 75% of the world's crude oil supply)

---

[2] Possible expansions of Hotelling's rule are suggested by e.g. Krautkraemer (1998) and Gaudet (2007).

than to the exhaustion of oil. The most significant price movements during the last century are associated with the first assertion of power by OPEC, the Arab oil embargo (1973), the first Gulf War (1990-1991) and the increase in commodity prices (around 2005). This increase was partly stopped by the Financial Crisis (2008). The supply cut by the OPEC (2009) and the Arab spring (2010) led to further price increases until mid 2014. The high oil prices and technological progress led to the increased utilization of "unconventional" sources, such as offshore oil, tar sands, tight oil, heavy oil, shale gas, coal-bed methane, shale oil, and oil shale. These oil sources are considered "unconventional" because previously the extraction process was economically not profitable. Recently hydraulic fracturing, or fracking has drastically changed the petroleum industry. Fracking is a technique for shooting water mixed with sand and chemicals into rock, splitting it open, and releasing previously inaccessible oil, referred to as tight oil. Fracking is criticized as an environmental menace to underground water supplies, and may eventually be greatly restricted. The increased accessibility of tight oils in North America would enable the United States to satisfy its own oil demand within a decade (at least for a while). Increased fracking led to a loss in market shares for Saudi-Arabia, the world largest oil producer. As a reaction Saudi-Arabia started to dump oil on the market to shake the weak production out of the market. This is one reason for the significant price decline until today.

Returning to the original question of what effect the exhaustion of natural resources has on economic growth; the pessimistic view on limited economic growth remains. Meadows et al. (1972) concluded that the long-run prospect may be worse than stagnation. They predicted the starvation of mankind in the near future. Data from 1900-1970 established a baseline of past behavior on which a model to forecast the events of the period 1970-2100 was based. Based upon the observed trends of resource use, the world should have run out of the main exhaustible resources (oil, gas, iron and copper) within a thirty years range. The result is depicted in figure 2.2.

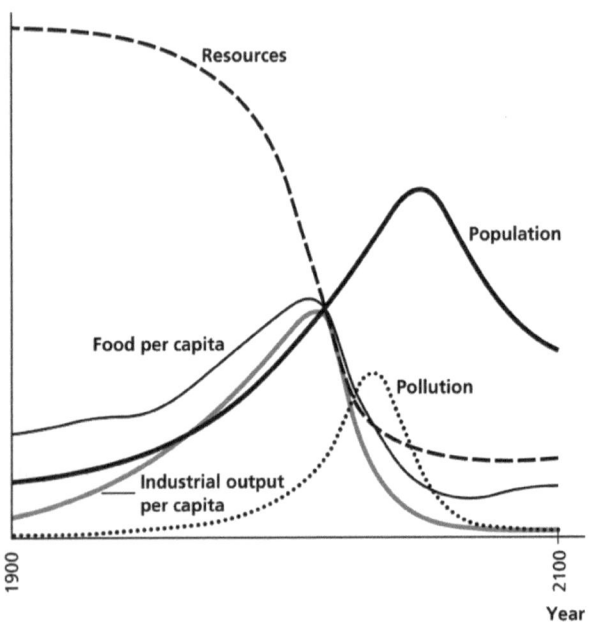

Fig. 2.2: Growth Forecast from the Limits to Growth.
Source: Meadows et al. (1972), Figure 35.

So far this doomsday scenario has been utterly wrong. Although the current level of resource consumption may not be sustainable, economic growth is not necessarily limited.

The evolution of the analysis process can be understood using the example of crude oil. For many years it has been the world's most important source of energy. It is used to generate heat, drive machinery, and fuel vehicles. Crude oil supplied about 33% of global energy needs in 2014 (followed by coal (30%), natural gas (24%), renewable resources (9%), and nuclear (4%)).[3] Its components are used to produce almost all chemical products, such as plastic, detergent, paint, and even

---

[3] Compare BP (2014, p. 40).

medicine. Oil affects all aspects of daily life. The percentage of market shares of oil associated with different aspects of daily life are depicted below.

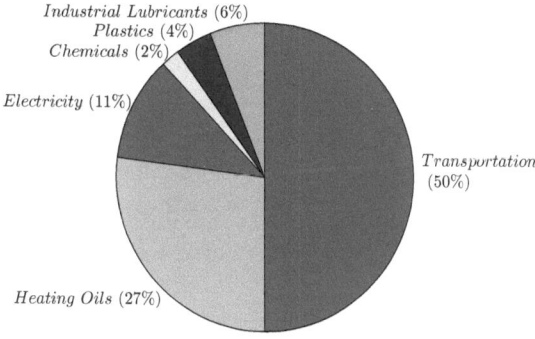

Fig. 2.3: Global Usage of Oil.
Source: National Oceanic and Atmospheric Administration (2011).

Oil is one of the world's largest traded commodities (measured in value or volume) due to its broad implementation range. Deutsche Bank (2013, p. 109) estimates that 59% of the commercially recoverable oil reserves have already been extracted and consumed. Weil (2013, p. 486) estimates that the remaining quantity of crude oil will last 61 years at the current rate of use. However, historically the exhaustion of raw materials has been continuously postponed despite a rising consumption rate. Two important explanations are newly discovered oil reserves and technological advancements. Over the last 40 years the oil consumption has been almost entirely covered by the discovery of oil reserves or technological advancements. This tendency is shown in the following graphic.

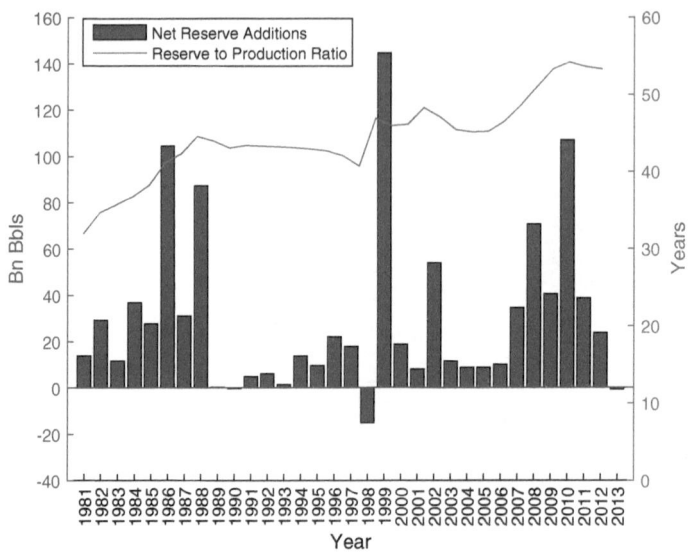

Fig. 2.4: Net Difference Between Annual Reserves Additions and Annual Consumption.
Source: Own illustration, based on data from BP (2014) Statistical Review of World
Energy.

Despite increased consumption the reserve to production ratio has been steadily
increasing. However, the oil supply is limited and with continuous consumption
will be exhausted. The question remains whether the world economy can persist
without considerable extraction of crude oil.

The negative effect of resource depletion has historically been compensated by
other factors leading to continued economic growth. Substitution is one of the most
basic factors. The exhaustible resource can be substituted by physical capital or
another natural resource (if available). Natural substitutes for oil are natural gas or
coal. In addition, a machine that uses oil more efficiently is an example of capital
substitution.

Another factor is recycling. The increasing use of exhaustible resources is
associated with increased waste. This offers new opportunities. The waste can be

recycled and used as a secondary material. Around 4% of world's oil production is used as a feedstock to make plastics.[4] The plastic market is a good example of how waste can be recycled thus reducing the total consumption of oil. In the European Union 62% of post-consumer plastics waste is recovered through recycling and energy recovery processes while 38% ends up on the waste dump.[5] Even though a lot of plastic is recycled worldwide, the effect on the depletion of oil is rather small. However, recycling may be a medium-term solution to continued growth. Additionally, recycling is interesting from an environmental point of view.

In the long-run technological progress is needed. Technological progress not only extends the economic lifetime but eases resource constraints that impede economic growth. Resource-augmenting technological progress can increase the efficiency of natural resource use, i.e. more output can be produced from a given amount of natural resources or less resources are needed for a given output. A specific example is the supplementation of fossil fuels with ethanol (E10).

Innovation reduces the cost of oil extraction from "unconventional" sources, increasing the lifetime of oil. A specific example is fracking. Moreover, technological progress can lead to an alternative technology which uses a different natural resource. New technology may allow oil to be replaced by methane hydrate, a crystalline natural gas located beneath the seafloor which exists in immense quantities. Methane hydrate is estimated to be twice as abundant as all other fossil fuels combined. However, all nonrenewable resource substitutes can only serve as temporary solutions. These "bridge fuels" will substitute oil until a technology is found which uses renewable resources and thus leads to sustained economic growth. A specific example is the use of alternative energy sources like sunlight, wind, water, and geothermal heat.

The thesis explains recycling and technological progress as a solution to the limited availability of fossil fuels using theoretical economic growth models.

---

[4] Compare PlasticsEurope (2014, p. 3).
[5] Compare PlasticsEurope (2015, p. 20).

3   The Dasgupta-Heal Model

Before introducing recycling and technological progress a simple economic growth model adapted from Dasgupta and Heal (1974) is introduced and analyzed. A stylized characterization of the model is given below:

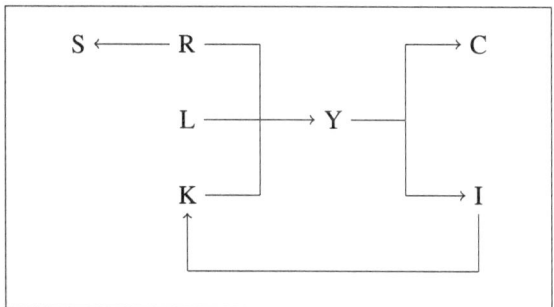

The economy is characterized by a given stock of an exhaustible natural resource S_0, which is continuously reduced by the flow R_t of resource extraction:

$$\dot{S} = -R_t. \tag{3.1}$$

In addition to the given natural resource stock, the economy starts with a given physical capital stock K_0. The depreciation rate of physical capital, e.g. stemming from wearing out, breaking down, or technological obsolescence of machinery, is negligible. Physical capital K_t, labor input L_t, and the flow of resource extraction R_t are used to produce the output Y_t of the economy, which is then either consumed or invested. Consumption increases current well-being but leads to a negative change in capital. Investment in physical capital stock leads to enhanced production:

$$\dot{K} = F(K_t, L_t, R_t) - C_t. \tag{3.2}$$

With enhanced production greater consumption in the future is possible.

At this point a simplification is possible. Although a logistic diffusion function could be used to model population growth many models describe the limited carrying capacity of the earth. A maximum population of approximately 13 billion has been predicted.[6] On a time-scale appropriate to finite resources it seems reasonable to argue that population is constant. The labor input is proportional to the growth rate of the population. With a constant population the growth rate is zero and labor input can be normalized to $L = 1$.

The production function $F(K_t, R_t)$ is assumed to be homogenous of degree one (it exhibits constant returns to scale ($\alpha_1 = 1 - \alpha_2 =: \alpha$)). α is the output elasticity of capital and $1 - \alpha$ the output elasticity of the natural resource. The marginal products of the production function are positive and diminishing for both inputs. A variation of the empirically found Cobb-Douglas production function (Cobb and Douglas, 1928) is used due to its mathematical simplicity:

$$Y = F(K_t, R_t) = K^\alpha R^{1-\alpha}. \tag{3.3}$$

The fulfillment of the Inada conditions by the Cobb-Douglas production function guarantees a stable economic growth path in the neoclassical growth model.[7]

In the words of Cass (1965, p. 234), "social welfare is related to the ability of the economy to provide consumption goods over time." The household's utility function $U(C_t)$ describes the development of welfare over time. The household's future utility is dependent on every descendent's utility. The utility function depends on consumption and exhibits positive but diminishing marginal utility. Cass (1965, p. 234) makes the case, "consumption tomorrow is not the same thing

---

[6]   Compare e.g. Cohen (1995).

[7]   Compare Inada (1963).

as consumption today." There is a stronger obligation to the present and near future than to the infinite future, thus a time preference rate ρ is implemented. The present value of utility is maximized for optimal social welfare:

$$\max_{C,R} \int_0^\infty e^{-\rho t} U(C_t) dt \tag{3.4}$$

subject to the dynamic capital accumulation, the process of resource extraction:

$$\dot{K} = F(K_t, R_t) - C_t$$
$$\dot{S} = -R_t$$

and the boundary and non-negativity constraints:

$$K(0) = K_0 \, , \; K(t) \geq 0$$
$$S(0) = S_0 \, , \; S(t) \geq 0.$$

The solution to the dynamic optimization problem uses the maximum principle of optimal control and combines the optimality conditions. The solution to this constrained optimization problem is derived in Appendix A. Using subscripts to denote partial derivatives and omitting the argument t, the first-order conditions are given by:

$$H_C: \; \psi_1 \overset{!}{=} e^{-\rho t} U'(C) \tag{3.5}$$
$$H_K: \; -\dot{\psi}_1 \overset{!}{=} \psi_1 F_K \tag{3.6}$$
$$H_R: \; \psi_2 \overset{!}{=} \psi_1 F_R \tag{3.7}$$
$$H_S: \; -\dot{\psi}_2 \overset{!}{=} 0. \tag{3.8}$$

These first-order conditions are evaluated further. It is socially optimal to completely deplete the stock of the exhaustible resource, equation 3.8. The optimal consumption and savings statement is given by equation 3.5. A marginal unit of output can be used for consumption purposes or for investment. An efficient outcome is one were both are equally beneficial, i.e. where the shadow price of capital ψ_1 equals the marginal utility. The left side of equations 3.6 and 3.7 is the marginal opportunity cost of using one more unit of capital or natural resource, respectively, and employing it in final production. The right side corresponds to the benefits. The increase of a single input of any production factor increases the level of output.

Combining the previously stated first-order conditions yields Ramsey's rule, which defines the optimal consumption path as:

$$\frac{\dot{C}}{C} = -[F_K - \rho]\frac{U'(C)}{U''(C)C}.$$

(3.9)

The rate of consumption along an optimal path depends on the time preference rate ρ, the inter-temporal elasticity of substitution $\gamma = -\frac{U'(C)}{U''(C)C}$, and the marginal product of physical capital F_K. The inter-temporal elasticity of substitution is commonly defined as the willingness of a household to substitute between future and present consumption. When choosing between consumption and saving, households consider the difference between the market interest rate (marginal product of capital) and their own "impatience". If the marginal product of capital (interest rate) exceeds the time preference rate then consumption today is lower than in the future, a positive growth rate is observed.

Once steady state has been reached, there is no change in a household's consumption rate. The marginal product of capital equals the time preference rate and thus is constant. The resulting asymptotically constant inter-temporal elasticity of substitution can be achieved using the CIES (constant inter-temporal elasticity of substitution) utility function:

$$U(C) = \begin{cases} \frac{C^{1-\theta}-1}{1-\theta} & \text{if } \theta > 0, \theta \neq 1 \\ ln(C) & \text{if } \theta = 1. \end{cases}$$

(3.10)

For this utility function the inter-temporal elasticity of substitution is given by $\gamma = 1/\theta$, where θ is a parameter which governs the elasticity of inter-temporal substitution. With increasing θ households are less willing to accept fluctuations in consumption, i.e. they smooth consumption over time.

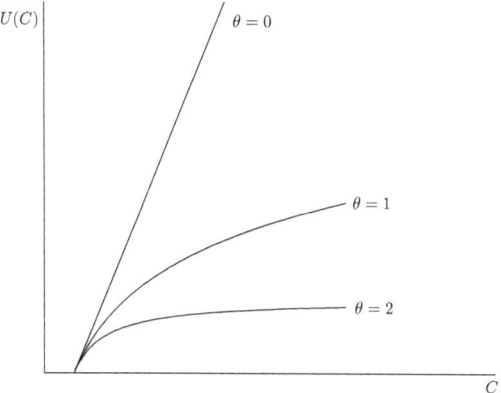

Fig. 3.1: CIES Utility Functions.
Source: Own illustration.

As θ approaches zero, the utility function becomes linear. Economically it can be understood that households are indifferent to current and future consumption. With increasing consumption the rate of utility declines faster with higher θ (corresponding to a low inter-temporal elasticity of substitution). The higher θ the less willing households are to accept deviations from a uniform consumption pattern over time.

There is empirical evidence that the inter-temporal elasticity of substitution is less than one.[8] Most estimates lie in the range of $\gamma \in (0.5, 1)$ implying $\theta \in (1, 2)$. Hence, it is plausible and convenient to assume $\theta = 1$, i.e. a logarithmic utility function.

---

[8] e.g. Havranek (2014) provides an overview and discussion of the estimated elasticities.

By inserting the production function and using the logarithmic utility function Ramsey's rule can be rewritten as:

$$\dot{C} = [\alpha K^{\alpha-1} R^{1-\alpha} - \rho]C. \tag{3.11}$$

Consumption can either grow, stay constant or decrease, depending on the time preference rate ρ and the marginal product of physical capital:

1. $\frac{\dot{C}}{C} > 0$ if $F_K > \rho$,
2. $\frac{\dot{C}}{C} = 0$ if $F_K = \rho$,
3. $\frac{\dot{C}}{C} < 0$ if $F_K < \rho$.

The optimal consumption path is depicted below.

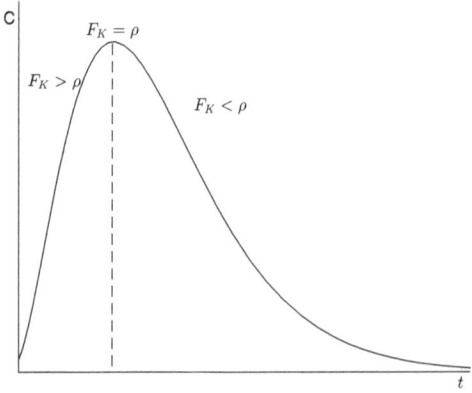

Fig. 3.2: Consumption Path.
Source: Own illustration.

Derived from the optimization problem, the growth rate of the marginal product of the natural resource equals that of physical capital (Hotelling rule):

$$\frac{\dot{F}_R}{F_R} = F_K. \tag{3.12}$$

In other words, along an optimal path the rate of return of physical capital (risk-free interest rates) equals the rate of return of the exhaustible resource.

For a production function homogenous of degree one, it is possible to write $x \equiv K/R$ and $f(x) \equiv F(K/R, 1)$, with x representing the capital to resource ratio. Using this definition, the elasticity of substitution between physical capital and the exhaustible resource is given by:

$$\sigma = -\frac{f'(x)[f(x) - xf'(x)]}{xf(x)f''(x)}. \tag{3.13}$$

Using the elasticity of substitution, the Hotelling rule can be rearranged to:

$$\frac{\dot{x}}{x} = \sigma \frac{f(x)}{x}. \tag{3.14}$$

How easily the natural resource can be substituted by physical capital plays an important role. The change of the capital to resource ratio equals the product of elasticity of substitution σ and the average product per unit of fixed capital $\frac{f(x)}{x}$. The average product per unit of fixed capital is an index for the importance of fixed physical capital in production. The higher the possibility of substitution and the higher the importance of physical capital in production is, the more natural resources are substituted by physical capital. In the case of the Cobb-Douglas production function, output is only possible with both input factors. Production is stopped when one of the inputs becomes zero. Hence, the substitution of the natural resource by capital is limited for a positive output.

Inserting the Cobb-Douglas production function into equation 3.14 leads to the following Bernoulli differential equation:

$$\dot{x} = x^{\alpha}. \tag{3.15}$$

The solution to this differential equation is the optimal time path of the capital to resource ratio given by:

$$x_t = [x_0^{1-\alpha} + (1-\alpha)t]^{1/(1-\alpha)}. \tag{3.16}$$

In terms of the capital to resource ratio the optimal consumption path can be restated as:

$$\frac{\dot{C}}{C} = \alpha x^{\alpha-1} - \rho \tag{3.17}$$

$$\frac{\dot{C}}{C} = \alpha [x_0^{1-\alpha} + (1-\alpha)t]^{-1} - \rho. \tag{3.18}$$

The capital to resource ratio x is monotonically increasing over time, due to the increasing substitution of the natural resource by physical capital. Since the production technology is dependent on both input factors, substitution of the natural resource by physical capital becomes increasingly impossible. The growth rate of the capital to resource ratio is declining.[9] Hence, it follows from Ramsey's rule that consumption converges to zero in the very long-run.[10]

The effect of capital accumulation on consumption is shown in the following graphic.

---

[9] For a formal proof see Appendix A.1.
[10] For a formal proof see Appendix A.1.

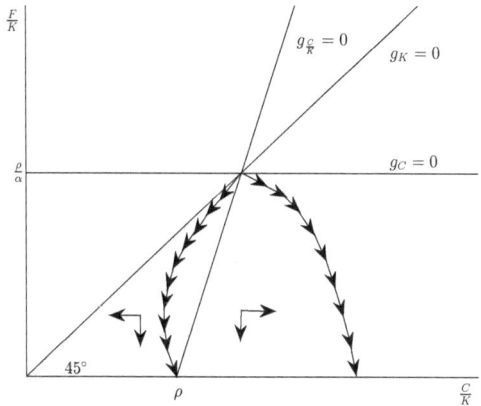

Fig. 3.3: Dasgupta-Heal: Phase Diagram in C/K-F/K Space.
Source: Own illustration.

The intersection of the upward sloping $g_{\frac{C}{K}} = 0$ line and the horizontal $g_{\frac{F}{K}} = 0$ is $(\rho, 0)$, the unique steady state point.[11] Note that the horizontal $g_{\frac{F}{K}} = 0$ line equals the C/K axis. In addition, the $g_C = 0$ and $g_K = 0$ loci are shown to characterize the transitional dynamics of consumption and capital accumulation. The locus $g_C = 0$ is an horizontal line at p/α. Above this line consumption increases and decreases below it. The locus $g_K = 0$ is a linear line with slope one. Capital increases above and decreases below the g_K-line.

The corresponding expressions of the constant asymptotic growth rates of K, Y and C, when the economy converges towards the steady state can be computed. Since $g_{\frac{F}{K}}$ and $g_{\frac{C}{K}}$ are both constant in steady state, $g_Y^* = g_K^* = g_C^*$ holds true. Using this and Ramsey's rule yields:

$$g_C^* = \alpha \frac{F}{K}^* - \rho \qquad (3.19)$$

$$g_C^* = g_K^* = g_Y^* = -\rho. \qquad (3.20)$$

---

[11] The derivation of this intersection point is given in Appendix A.2.

The ultimate result of the model is the convergence of capital, consumption, and production to zero and the starvation of mankind.[12]

3.1 A Closed-Form Solution

Pezzey and Withagen (1998) and Hartwick et al. (2003) showed that a closed-form solution of the Dasgupta-Heal model can be derived by assuming that the intertemporal elasticity of substitution equals the output elasticity of capital ($1/\theta = 1/\alpha$). There is no economic relationship between these two distinct parameters. $\theta = 1 = \alpha$ would imply that only capital is used for production. The closed-form solution is derived for $1/\theta = 1/\alpha \neq 1$, i.e. $0 < \alpha < 1$. The following utility function is considered:

$$U(C) = \frac{C^{1-\theta} - 1}{1 - \theta}. \tag{3.21}$$

The full solution is given in Appendix B, leading to the optimal time paths of the model's variables:

$$S_t = S_0 e^{-(\rho/\alpha)t} \tag{3.22}$$

$$R_t = \left(\frac{\rho}{\alpha}\right) S_0 e^{-(\rho/\alpha)t} \tag{3.23}$$

$$K_t = \left(\frac{\rho}{\alpha}\right) S_0 e^{-(\rho/\alpha)t} \left[\left(\frac{\alpha K_0}{\rho S_0}\right)^{1-\alpha} + (1-\alpha)t\right]^{1/(1-\alpha)} \tag{3.24}$$

$$F_t = \left(\frac{\rho}{\alpha}\right) S_0 e^{-(\rho/\alpha)t} \left[\left(\frac{\alpha K_0}{\rho S_0}\right)^{1-\alpha} + (1-\alpha)t\right]^{\alpha/(1-\alpha)} \tag{3.25}$$

$$C_t = \left(\frac{\rho}{\alpha}\right)^2 S_0 e^{-(\rho/\alpha)t} \left[\left(\frac{\alpha K_0}{\rho S_0}\right)^{1-\alpha} + (1-\alpha)t\right]^{1/(1-\alpha)}. \tag{3.26}$$

---

[12] Assuming a positive time preference rate.

The optimal paths are depicted below.

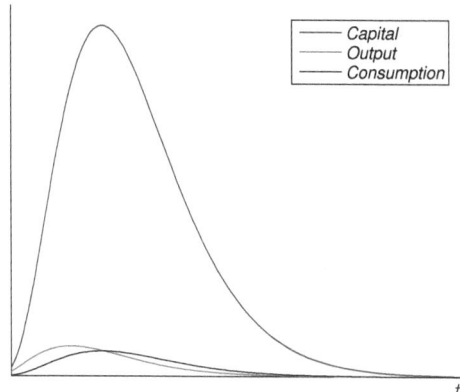

Fig. 3.4: Behavior of Output, Capital, and Consumption.
Source: Own illustration.

Consumption peaks at the same time as capital, whereas production peaks earlier.[13] Capital, consumption, and production converge to zero in the very long-run, leading to the starvation of mankind. The economy cannot survive.

3.2 Critical Assessment

The Dasgupta-Heal model serves as a starting point for an economic analysis of nonrenewable resources and long-term economic growth. Using the made assumptions the model results in the starvation of mankind. This prediction may be too pessimistic. The extremity of the prediction may be the result of the made assumptions. Many are simplifications and considerably abstract from reality.

---

[13] This is formally proved in Appendix B.

Physical capital consumption is possible but depreciation is neglected. In reality physical capital is depreciated. For example, machines wear out, break down, or technology may become obsolete. A more accurate prediction could be attained from the model through the introduction of a positive depreciation rate. This would ultimately lead to the same result, although the starvation of mankind would occur earlier. As previously explained it is assumed that the population and thus the labor force is constant. Moreover it is assumed, that the resource has a fixed nonrenewable stock. Additions to reserves forestall resource exhaustion, but would not fundamentally change the prediction of the model.

The Cobb-Douglas production function itself is an approximation of reality. Another possible production function is the constant elasticity of substitution (CES) production function. This function allows for a higher elasticity of substitution between the two input factors and production without the natural resource. Implementation of these production functions could avoid the starvation of mankind. The Cobb-Douglas function, however, serves as a good basis when considering natural resources, e.g. oil, because production is only possible with the resource. Additionally, it has been proven accurate for long-term economic forecasts.[14]

There is a dispute between economists about the time preference rate ρ.[15] According to Laibson (1997) a variable time preference rate should be used, considering the extreme impatience of households with regard to the immediate future and the higher patience about choices in the extended future. Ramsey (1928) and Rawls (1971) say that the time preference rate is unethical. Ramsey (1928, p. 543) prefers the use of a zero time preference, since in his opinion households "do not discount later enjoyments in comparison with earlier ones." According to Rawl's ethical principle, the outcome of the least well-of member of society is maximized. Solow (1974) applies this principle to natural resource economics. He shows that an equally distributed consumption over all generations is feasible if the output share for capital is greater than the output share for the natural resource in

---

[14] For a more detailed examination see e.g. Miller (2008).

[15] Compare e.g. Barro and Sala-i Martin (2004, p. 140f.) or Perman et al. (2003, Ch. 3) for a more detailed discussion on this topic.

the case of a Cobb-Douglas production function with no population growth and no technological progress. In this case capital accumulation offsets the resource depletion. In particular, the rents from nonrenewable resource extraction are invested into capital accumulation at each point in time.[16] Authors who prefer the use of a zero time preference rate argue that while individuals might be myopic, in social-decision making every generation must be treated equally. In the words of Heijman (1991, p. 82) "[as] sympathetic this kind of reasoning is, it misses the point. When we talk in economic terms, the question is not whether it is ethically right to discount future wealth or whether it is rational for us to do so. The fact is that future wealth is discounted on the individual scale as well on the social one. An economist has to examine the consequences of this behavior whether he agrees with it or not."

Two important factors that the basic model does not consider are recycling and technological advancement. Recycling can increase the lifetime of an economy through secondary material input and thus should be incorporated. Technological advancement can lead to more efficient productivity or to a sustainable substitute. In modern society as resources have become less available recycling and efficient use (through technological advancement) play a significant role. In the following the basic model is expanded to include these factors.

---

[16] Compare Hartwick (1977).

4 Recycling as a Source of Regeneration

The effect of recycling on resource conservation depends on the percentage of materials that economically and technically may be recycled. Recycling extends the lifetime of an economy since it increases the long-term supply of an exhaustible resources by a factor of its "recycling multiplier" as Tietenberg and Lewis (2012, p. 184) note.

There is extensive literature on the effects of recycling on economic growth.[17] In the early seventies the waste management was seen as "a spatially circumscribed problem of limited relevance, with no consequences for the economy as a whole" (Di Vita, 2005, p. 159). Initially, economists considered recycling from a microeconomic point of view. Recently, increased global waste production has resulted in research from a macroeconomic point of view.

Research is divided into two sectors, analysis of the conservation of nonrenewable resources and analysis of the alleviation of waste disposal problems (i.e. recycling as a pollution abatement activity). The conservation of nonrenewable resources is considered in more detail.

Weinstein and Zeckhauser (1974) and Schulze (1974) were among the first to study the conservation of nonrenewable resources. They theorize the optimal consumption pattern of exhaustible resources with recycling. Weinstein and Zeckhauser (1974) allow for complete recycling whereas Schulze (1974) considers the more realistic case where only an exogenous given share of waste is recyclable. Schulze (1974) includes a natural decay rate of the waste stock in his analysis, i.e.

---

[17] Compare e.g. Wacker (1987) for an overview of the literature of the seventies.

a bio-decomposition parameter. Kemp and Long (1980) make these two parameters endogenous, by introducing labor effort into the recycling process.

In addition, Mäler (1974) and Wacker (1987) analyze the utilization of recycled resources versus virgin resources. They show that even if the input substitute (recycled resource) is more expensive than the primary production factor, it will always be produced before the cheapest exhaustible resource is exploited, since waste recycling extends the lifetime of the economy. Recycling can partially offset the limits to growth stemming from the increasing scarcity of the exhaustible resources.

More recently, numerous economists integrated recycling into an endogenous growth theory introduced by Romer (1990), Lucas (1988), Grossman and Helpman (1991), and Rebelo (1991) under a material balance constraint.[18] Examples are Di Vita (2001), Kuhn et al. (2003), Di Vita (2006, 2007) and Pittel et al. (2006). Di Vita (2001) deals with exhaustible resources and shows that technological progress plays a crucial role in increasing the quantity of secondary materials. Di Vita (2006) extends this result to renewable resources. In both articles, recycled and virgin stocks are assumed to be perfect substitutes. In contrast, Di Vita (2007) considers the case of imperfect substitutes.

Pittel et al. (2006) focus on the implications of recycling-related market failures and show that optimal environmental policy can serve as a means to raise economic growth. Pittel et al. (2010) use the material balance constraint in a neoclassical growth framework. They show that the stocks of virgin resources and those of recycled resources approach zero in the long-run. A detailed description of this model is given later.

---

[18] The material balance constraint is discussed in more detail later on.

4.1 The Economic System and the Environment

As previously shown, economic growth is limited by the finite amount of natural resources. This limitation allows for an alternative view on the optimal depletion of exhaustible resources. Nature is no longer seen as part of the economy, but the economy is seen as part of nature. The economy is an open subsystem of the larger but finite, closed, and non-growing system of nature. The dynamics of economy-environment interactions appear as a co-evolution of two systems, each with internal structure and dynamics. The environment and the economy mutually influence each other's development. Ayres and Kneese (1969) and Kneese et al. (1970) were among the first to incorporate the law of conservation of mass (referred to as the material balance principle) into the analysis of economic growth. The material balance principle states that matter can neither be created nor destroyed but transformed.[19] In other words, all natural resources that enter the economic system must always be accounted for, i.e. "durable goods, recycled inputs, or waste products deposited into the air, land, or water" (Kolstad and Krautkraemer, 1993, p. 1221). The material balance principle constrains economic production possibilities. For example, materials not used to create durable goods or recycled inputs must return to the environment as waste. This results in the need for the improved durability of products, capital accumulation, and perhaps most importantly, the recovery and recycling of materials. A graphical description is given in figure 4.1.

---

[19] Note in the equivalent formulation of the first law of thermodynamics energy is used instead of matter.

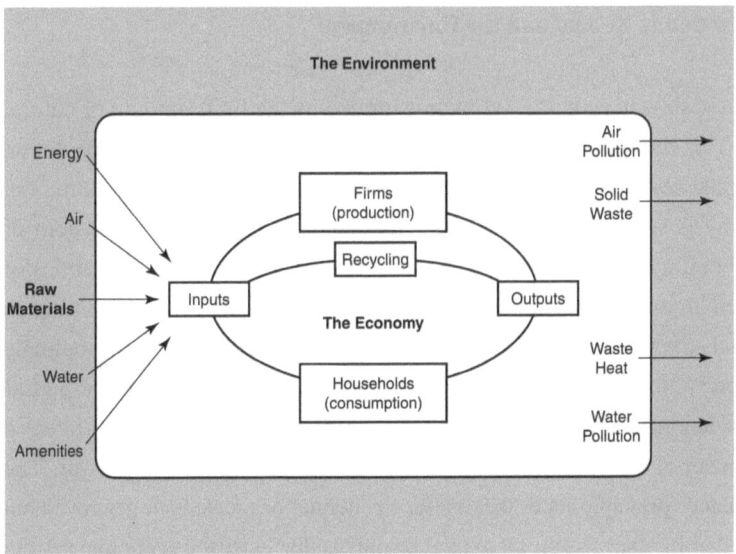

Fig. 4.1: The Economic System and the Environment.
Source: Tietenberg and Lewis (2012, p. 17).

In the words of Tietenberg and Lewis (2012, p. 18), this model can be applied because "historically speaking, for material inputs and outputs (not including energy), this system [earth] can be treated as a closed system because the amount of exports (such as abandoned space vehicles) and imports (e.g. moon rock) are negligible." It is important to keep in mind that this is only an approximation. In reality the earth as an environment is clearly not a closed system. Most energy is derived from the sun, either directly or indirectly.

4.2 Recycling Under a Material Balance Constraint

Pittel et al. (2010) extended the Dasgupta-Heal model using the material balance principle by introducing recycled waste as an input factor of production. The focus of the model lays on the conservation of nonrenewable resources. Possible negative effects on the environment are disregarded. A stylized characterization of the model (referred to as recycling model) is given below:

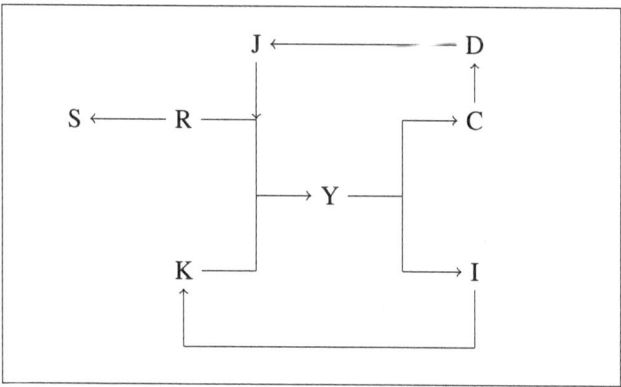

The economy is characterized by a given stock of an exhaustible resource S_0, which is continuously reduced by the flow R_t of resource extraction of the virgin resource:

$$\dot{S} = -R_t. \tag{4.1}$$

The extraction costs of the virgin resource as well as recycling costs are neglected. In addition to the given natural resource stock, the economy starts with a given physical capital stock K_0. The depreciation rate of physical capital is negligible. Secondary material input from recycled waste J_t is considered an additional input factor. Physical capital K_t, virgin resource input R_t, and secondary material inputs from recycled waste J_t are used to produce the output Y_t of the economy. The output is then either consumed or invested. Investment in physical capital increases its stock:

$$\dot{K} = F(K_t, R_t, J_t) - C_t. \tag{4.2}$$

With enhanced production greater consumption in the future is possible. The Cobb-Douglas type production function is given by:[20]

$$Y = F(K, R, J) = K^\alpha R^\beta J^\phi. \tag{4.3}$$

α is the output elasticity of capital. β and ϕ are the output elasticity of the virgin and the recycled resource respectively. It holds true that $\alpha + \beta + \phi = 1$. A specific example in the case of oil is the production and recycling of PET bottles. PET bottles are produced from fossil fuels. The recycled bottles are used as raw materials for various products, e.g. for polyester fibers (a base material for the production of clothing, pillows, carpets, etc.).

All materials that are discarded after consumption accumulate to waste stock D_t which is recycled and reused.

$$\dot{D} = -J_t + (J_t + R_t)\kappa. \tag{4.4}$$

$\kappa_t = \frac{C_t}{F(K_t, R_t, J_t)}$ denotes the share of the natural materials that ends up on the waste dump.[21] The waste stock D_t decreases with the amount of materials taken from the waste dump for recycling J_t and increases with the share of the natural materials that ends up on the waste dump κ_t. This share is determined by the inter-temporal consumer preferences.

As previously stated, under the material balance principle all natural resources that enter the economic system must always be accounted for. Note, that this

---

[20] In contrast to Pittel et al. (2010) technological progress is not included, i.e. the level of technology A is normalized to one.

[21] In the present model this is identical to the consumption share of output, since capital and consumption goods are produced by the same technology.

implies the use of a double system of units. While the quantities of the virgin and recycled resources are measured in physical units (atomic mass), capital, output, and consumption are measured in the number of units used or produced. In the present model the materials must be accounted for post consumption. This can be understood since "a share of materials becomes bound in the capital stock while the rest is completely discarded onto a waste pile after consumption" (Pittel et al., 2010, p. 382).

Natural degeneration is neglected. Moreover, there are no restrictions on recycling, i.e. there is enough energy and storage space for complete recycling.[22]

The present value of utility is maximized for optimal social welfare:

$$\max_{C,R,J} \int_0^\infty e^{-\rho t} U(C_t) dt \qquad (4.5)$$

subject to the dynamic capital accumulation, the process of virgin resource extraction, the process of waste recycling:

$$\dot{K} = F(K_t, R_t, J_t) - C_t$$
$$\dot{S} = -R_t$$
$$\dot{D} = -J_t + (R_t + J_t)\kappa_t$$

and the boundary and non-negativity constraints:

$$K(0) = K_0 \,,\ K(t) \geq 0$$
$$S(0) = S_0 \,,\ S(t) \geq 0$$
$$D(0) = D_0 \,,\ D(t) \geq 0.$$

The solution to the dynamic optimization problem uses the maximum principle of optimal control and combines the optimality conditions. The solution to this constrained optimization problem is derived in Appendix C.

---

[22] Ayres (1999) examines this topic in further detail.

Using subscripts to denote partial derivatives and omitting the argument t, the first-order conditions are given by:

$$H_C: \quad \psi_1 \quad \overset{!}{=} e^{-\rho t} U'(C) + \psi_3 \mu \tag{4.6}$$

$$H_K: \quad -\dot{\psi}_1 \quad \overset{!}{=} \psi_1 F_K - \psi_3 \mu \kappa F_K \tag{4.7}$$

$$H_R: \psi_2 - \psi_3 \kappa \overset{!}{=} \psi_1 F_R - \psi_3 \mu \kappa F_R \tag{4.8}$$

$$H_S: \quad -\dot{\psi}_2 \quad \overset{!}{=} 0 \tag{4.9}$$

$$H_J: \psi_3 - \psi_3 \kappa \overset{!}{=} \psi_1 F_J - \psi_3 \mu \kappa F_J \tag{4.10}$$

$$H_D: \quad -\dot{\psi}_3 \quad \overset{!}{=} 0. \tag{4.11}$$

$\mu = \frac{R+J}{F}$ is the material intensity of output. These first-order conditions are evaluated further. Similar to the Dasgupta-Heal model, it is socially optimal to completely deplete the stock of the exhaustible (virgin) resource, equation 4.9. In the recycling model the same holds true for the recycled resource, equation 4.11. The optimal consumption and savings statement depicted in equation 4.6 is extended using a correction term. The shadow price of capital equals the marginal utility and "the potential marginal Recycling Value as a by-product of Consumption" (Pittel et al., 2010, p. 384). Recycling produces a consumable good from waste. This re-useability increases the value of an output unit of consumption.

In equations 4.8 and 4.10 the left side is the marginal opportunity costs of extracting one more unit of a virgin or recycled resource, respectively, and employing it in the final output production. The marginal opportunity costs of resource extraction and/or waste recycling are reduced by the "rental" element $\psi_3 \kappa$ in the pricing of virgin resources and recycled materials. This is due to the re-useability of natural resources. κ is the share of each unit of extracted or recycled materials that can be used again in future production. The right side corresponds to the benefits of extracting one marginal unit of virgin or recycled resources. The increase of a single input of any production factor increases the level of output not only directly (first term), but also indirectly through the generation of valuable waste (second term). The second term can be interpreted as "the potential marginal Recycling Value of the input [factors] (...) in Production" (Pittel et al., 2010, p. 384).

Rearranging equation 4.10 makes this relationship very clear:

$$\psi_3(1 - \kappa) \stackrel{!}{=} F_J[(1 - \kappa)\psi_1 + \kappa e^{-\rho t}U'(C). \tag{4.12}$$

The total costs of extracting one marginal unit from the waste dump and using it in production equals the benefits. From the marginal additional output F_J only κ is consumed and recycled while $1 - \kappa$ is saved, i.e. it ends up in capital accumulation. Evaluating the shares by the shadow price of capital and marginal utility of consumption respectively, gives the benefits of extracting one marginal unit of waste.

Combining the previously stated first order conditions yields the following results. The Hotelling rule for recycled waste is given by:

$$\frac{\dot{F_J}}{F_J} = F_K + \left[\frac{F_J}{1 - \kappa}(\dot{\mu}\kappa + \dot{\kappa}\mu) - \frac{\dot{\kappa}}{1 - \kappa}\right]. \tag{4.13}$$

The Hotelling rule for virgin resources is:

$$\frac{\dot{F_R}}{F_R} = F_K + \left[\frac{F_J}{1 - \kappa}(\dot{\mu}\kappa + \dot{\kappa}\mu) - \frac{\dot{\kappa}}{1 - \kappa}\frac{F_J}{F_R}\right]. \tag{4.14}$$

The standard Hotelling rule (see equation 3.12) is enhanced by a composite term that captures the circulation effect of recycling. The recycling of an additional marginal unit induces a change of the share of materials flowing back after consumption, $\dot{\kappa}\mu$. This affects the future producible output. In addition, a change in the share of natural materials that end up on the waste dump $\dot{\kappa}$ directly affects the future availability of materials for recycling. If in the future more materials end up on the waste dump (positive $\dot{\kappa}$) the opportunity costs of extraction decrease.

Derived from the optimization problem, the optimal consumption path (Ramsey rule) is given by:

$$\frac{\dot{C}}{C} = F_K - \rho + \frac{F_J \mu}{(1-\kappa)(1-F_J\mu)}\left[F_K(1-\kappa) + \frac{\dot{\mu}}{\mu}\right]. \qquad (4.15)$$

An additional term is present in the optimal consumption path versus the Dasgupta-Heal model (see equation 3.11). The term reflects the effects of introducing recycling and the material balance principle. Along an optimal path the consumption rate is positive as long as the marginal product of capital plus the additional term exceeds the time preference rate. Due to recycling the lifetime of the economy is extended. However, in the long-run consumption approaches zero:[23]

$$g_Y^* = -\rho. \qquad (4.16)$$

Although a reuse of virgin as well as of recycled resource is possible, the overall stock of natural resources is limited. Only a fraction of the output is consumed and thus recycled. The rest is bound in capital. The amount of waste available for recycling decreases in the long-run. The natural resources are increasingly substituted by capital. Once the virgin resource is used up, the recycled resource becomes increasingly scarce. Production is stopped. The remaining capital stock is consumed. The economy will not survive.

4.3 Incomplete Recycling

Under the material balance principle, the laws of thermodynamics states that with enough energy all transformations of matter are possible. Theoretically complete recycling is possible. Assuming complete consumption and no capital accumulation would allow for infinite growth. The resource shortage would no longer be a

---

[23] Compare Appendix C.1.

constraint on economic growth. This is at odds with reality. In case of oil, or more specifically in the case of PET bottles, complete recycling is technically impossible, due to a fractional loss of material (mass) through use. In other words, only a share of waste from consumption returns to the waste pile while the other fraction is impossible to recover. The model is now modified to consider imperfect recycling. The motion of waste is then given by:

$$\dot{D} = -J + (J + R)(1 - \chi)\kappa. \qquad (4.17)$$

where $0 < \chi < 1$ denotes the fractional loss of material through use. The optimal consumption path is:

$$\frac{\dot{C}}{C} = F_K - \rho + \frac{F_J \mu}{\frac{1}{1-\chi} - \kappa + F_J \mu \kappa - \mu F_J} \left[F_K(1 - \kappa) + \frac{\dot{\mu}}{\mu} \right]. \qquad (4.18)$$

The previously obtained results are still accurate. Consumption, however, peaks earlier. Since only part of the consumed output can be recycled, the amount of waste available for recycling is more rapidly used. The lifetime of the economy will be shorter compared to complete recycling.

5 Technological Progress

While recycling can extend the life of the economy and is regarded as an intermediate solution, technological progress is needed to sustain long-term economic growth.

There is extensive literature on exhaustible resources including technological progress.[24] One of the first models for the optimal extraction of exhaustible resources, considering exogenous technological progress was formulated by Stiglitz (1974). The production function is extended with a technological progress parameter, without specifying whether technological change is capital- or resource-augmenting. This extension prevents the starvation of mankind. This model will be described in more detail later on.

In the subsequent chapters a backstop technology is introduced. According to Nordhaus et al. (1973) a backstop technology is an ultimate production technology which uses an abundant (infinite) resource and capital as inputs. In subsequent articles, several economists have included backstop inputs in models of exhaustible resources. Examples are Dasgupta and Heal (1974) and Kamien and Schwartz (1978). Both are discussed in more detail later on. In these two models, the backstop technology delivers a given stream of utility, or is a perfect substitute for the exhaustible resource in production. The arrival date of the backstop technology is uncertain. In Dasgupta and Heal (1974) the backstop technology is not a smooth gradual innovation process but rather a technological breakthrough arriving in a discrete once-for-all manner with economy-wide consequences. The arrival date of

---

[24] Compare e.g. Krautkraemer (1998) and Groth (2006) for an overview.

the backstop technology is exogenously given. In Kamien and Schwartz (1978), the arrival date is determined endogenously by investments in *R&D*. Nevertheless, as soon as the backstop input has arrived, the exhaustible resource is abandoned.

Tahvonen and Salo (2001) depart from the discrete once-for all manner and consider an economy where nonrenewable and renewable resources coexist and are perfect substitutes. In this case both resources are used during a transition period, while in the long-run the exhaustible resource is exhausted. Tsur and Zemel (2003, 2005) also consider a continuous improvement of an existing backstop technology. The exhaustible resource and the backstop technology exist since the beginning of time. They are perfect substitutes. The cost of using the backstop technology can be gradually reduced by investment in *R&D*. Similar to Tahvonen and Salo (2001), they find that the exhaustible resource and the backstop technology are used simultaneously until the nonrenewable resource is depleted.

Just et al. (2005) consider the case where a backstop technology capable of substituting the resource already exists but the discovery of a superior technology is expected. The discovery date of the superior technology is uncertain. They find that it may be favorable to wait for the superior backstop technology instead of using the already existing backstop technology.

Additionally, the endogenous growth theory introduced by Romer (1990), Lucas (1988), Grossman and Helpman (1991), and Rebelo (1991) has been applied to exhaustible resources. Examples are Bretschger (1998), Barbier (1999), Schou (2000), Acemoglu (2002), Bretschger and Smulders (2004), Grimaud and Rougé (2003, 2005), Acemoglu and Aghion (2012), and Smulders and Withagen (2012). A detailed analysis of this theory exceeds the realm of this thesis.

5.1 Exogenous Technological Progress

A first attempt to answer the question whether technological progress is able to prevent the starvation of mankind was done by Stiglitz (1974). The characterization of the economy is similar to the Dasgupta-Heal model, described in chapter 3. In short, a given stock of an exhaustible natural resource S_0 is continuously reduced through the flow R_t of resource extraction:

$$\dot{S} = -R_t. \tag{5.1}$$

In addition to the given natural resource stock, the economy starts with a given physical capital stock K_0. Physical capital K_t and the flow of resource extraction R_t are used to produce the output Y_t of the economy, which is then either consumed or invested. In contrast to the basic Dasgupta-Heal model, the production function is extended with a resource-augmenting technological progress parameter A:

$$F(K,R) = K^\alpha (AR)^{1-\alpha}. \tag{5.2}$$

The technological level $A = A_0 e^{\delta t}$ grows at the constant and exogenous given rate δ.[25] Resource-augmenting technological progress increases the efficiency of natural resource use, i.e. more output can be produced from a given amount of natural resources or less resources are needed for a given output. A specific example is the supplementation of fossil fuels with ethanol (E10).

Investment in physical capital stock leads to enhanced production:

$$\dot{K} = F(K_t, A_t R_t) - C_t. \tag{5.3}$$

With enhanced production greater consumption in the future is possible.

---

[25] The specified production function is equivalent to the one specified in Stiglitz (1974): $Y = e^{\lambda t} K^\alpha R^{1-\alpha}$ with $\delta = \frac{\lambda}{1-\alpha}$ and $A_0^{1-\alpha} = 1$.

For the purpose of simplicity, efficiency units, i.e. $k = K/A$ and $y = Y/A = \tilde{F}(k, R)$, are used to describe capital and output. The side condition for capital accumulation in efficiency units is:

$$\dot{k} = \tilde{F}(k_t, R_t) - C_t e^{-\delta t} - \delta k_t. \tag{5.4}$$

The present value of utility is maximized for optimal social welfare:

$$\max_{C,R} \int_0^\infty e^{-\rho t} U(C_t) dt \tag{5.5}$$

subject to the dynamic capital accumulation in efficiency units, the process of resource extraction:

$$\dot{k} = \tilde{F}(k_t, R_t) - C_t e^{-\delta t} - \delta k_t$$
$$\dot{S} = -R_t$$

and the boundary and non-negativity constraints:

$$k(0) = k_0 \ , \ k(t) \geq 0$$
$$S(0) = S_0 \ , \ S(t) \geq 0.$$

The solution to the dynamic optimization problem uses the maximum principle of optimal control and combines the optimality conditions. The solution to this constrained optimization problem is derived in Appendix D. Using subscripts to denote partial derivatives and omitting the argument t, the first-order conditions are given by:

$$H_C : \ \psi_1 e^{-\delta t} \overset{!}{=} e^{-\rho t} U'(C) \tag{5.6}$$
$$H_k : \ -\dot{\psi}_1 \overset{!}{=} \psi_1 \tilde{F}_k - \delta \tag{5.7}$$
$$H_R : \ \ \psi_2 \overset{!}{=} \psi_1 \tilde{F}_R \tag{5.8}$$
$$H_S : \ -\dot{\psi}_2 \overset{!}{=} 0. \tag{5.9}$$

Combining the previously stated first-order conditions yields Ramsey's rule, which defines the optimal consumption path as:

$$\frac{\dot{C}}{C} = [\alpha k^{\alpha-1}(R)^{1-\alpha} - \rho]. \tag{5.10}$$

Along an optimal path the consumption rate is positive as long as the marginal product of capital in efficiency units exceeds the time preference rate ρ. In the long-run consumption is only feasible if at least $\tilde{F}_k = \rho$ holds true. As a result, if t approaches infinity, $d\tilde{F}_k/dt$ must be zero, yielding:

$$\lim_{t\to\infty} \frac{\dot{\tilde{F}_k}}{\tilde{F}_k} \overset{!}{=} 0 \tag{5.11}$$

$$\frac{\dot{R}}{R} = \frac{\dot{K}}{K} - \frac{\dot{A}}{A}. \tag{5.12}$$

Exhaustible resources are diminishing over time. In the basic model from chapter 3, the diminishing resource is substituted by capital to maintain consumption. In the long-run the whole capital stock is consumed, leading to the starvation of mankind. This problem is addressed by introducing the resource-augmenting technology. Now capital is no longer entirely consumed. The rate of capital accumulation increases as long as it is less than the resource-augmenting technology growth rate. Otherwise, current consumption eventually provides greater present value utility than the future output of additional capital and capital accumulation stops.

In addition, along an optimal path the growth rate of the marginal product of the natural resource in efficiency units grows until the marginal product of physical capital in efficiency units equals the growth rate of the technological change (modified Hotelling rule):

$$\frac{\dot{\tilde{F}_R}}{\tilde{F}_R} = \tilde{F}_k - \delta \tag{5.13}$$

$$\alpha\frac{\dot{k}}{k} - \alpha\frac{\dot{R}}{R} = \alpha k^{a-1}R^{1-\alpha} - \delta. \tag{5.14}$$

Note that the marginal product $\tilde{F}_R$ increases with increasing k and decreases with increasing R. The opposite is true for $\tilde{F}_k$. For an increasing marginal product of the exhaustible resource, i.e. $\frac{\dot{\tilde{F}}_R}{\tilde{F}_R} = \tilde{F}_k - \delta > 0$, the ratio of capital in efficiency units k to resource input R increases. At the same time the marginal product $\tilde{F}_k$ decreases, leading to a lower growth rate of the marginal product of the natural resource in efficiency units. For $\frac{\dot{\tilde{F}}_R}{\tilde{F}_R} = \tilde{F}_k - \delta < 0$ the opposite is true, the k/R ratio decreases and $\tilde{F}_k$ increases. The growth rate of the marginal productivity of the natural resource in efficiency units increases. In the long-run $\tilde{F}_k = \delta$ holds true, implying a constant growth rate of the marginal product of the natural resource in efficiency units.

Using the previously obtained results, consumption can only increase or stay constant if $\delta \geq \rho$ holds true, i.e. the growth rate of the technological change is greater or equal to the time preference rate. Economically, a sufficiently high speed of technical progress δ is needed. The progress must counteract the impatience rate ρ which through the depletion process of the exhaustible resource leads to the starvation of mankind. Hence, if this condition is true the economy will survive.

To derive the steady state solution, the capital stock and the output in efficiency units adjusted by the resource input are considered ($\tilde{k} = K/AR$ and $\tilde{y} = F/AR$). Using this definition, the growth rates $g_{\frac{\tilde{F}}{\tilde{K}}}, g_{\frac{\tilde{F}}{\tilde{K}}}, g_V, g_C, g_K$ and g_R are calculated:[26]

$$g_{\frac{\tilde{F}}{\tilde{K}}} = -(1-\alpha)(g_K - g_R - \delta) \qquad (5.15)$$

$$g_{\frac{C}{\tilde{K}}} = g_C - g_K \qquad (5.16)$$

$$g_V = g_R + V. \qquad (5.17)$$

Note that V is the ratio of resource utilization (R) to the stock of the resource (S). In other words V is the flow to stock ratio.

The output to capital ratio decreases with increasing capital and increases with higher use of the exhaustible resource and with technological progress. The smaller the productivity of capital (α), the higher the influence of the growth rate of capital,

---

[26] Compare Appendix D.1.

the use of exhaustible resources and technological progress on $g_{\frac{F}{K}}$. The consumption to capital ratio increases when the growth rate of consumption exceeds the growth rate of capital. The flow to stock ratio increases when the growth rate of the resource use exceeds the negative resource utilization rate (-V).

The optimal growth path is depicted in the following phase diagram.

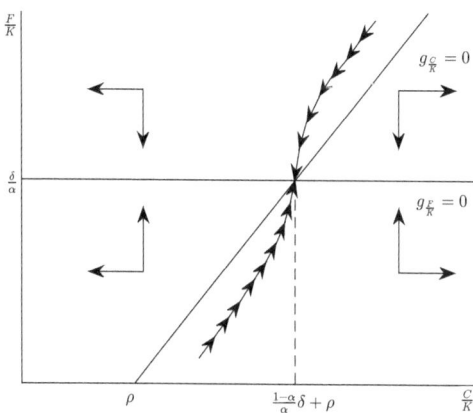

Fig. 5.1: Exogenous Technological Progress: Phase Diagram in C/K-F/K Space.
Source: Own illustration.

The intersection of the upward sloping $g_{\frac{C}{K}} = 0$ line and the horizontal $g_{\frac{F}{K}} = 0$ is $(\delta\frac{1-\alpha}{\alpha} + \rho, \frac{\delta}{\alpha})$, the unique steady state point. The optimal path is shown by the arrows towards the intersection point. Above (below) the $g_{\frac{C}{K}} = 0$ locus $\frac{C}{K}$ decreases (increases). The same holds true for the $g_{\frac{F}{K}} = 0$ locus. $\frac{F}{K}$ decreases above and increases below.

Assuming the economy follows the optimal path in figure 5.1, the optimal path in the $\frac{C}{K} - V$ space is depicted below.

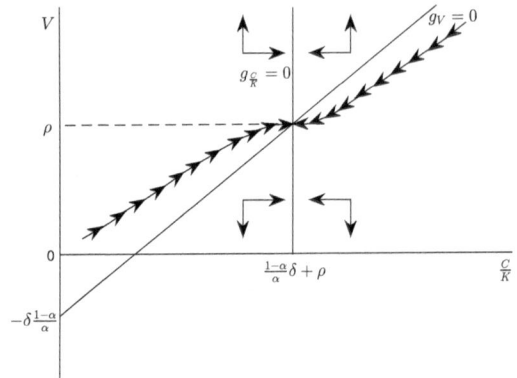

Fig. 5.2: Exogenous Technological Progress: Phase Diagram in C/K-V Space.
　　　　Source: Own illustration.

If the economy starts with a relatively high capital and low use of resources, then the lower optimal paths are considered. This is increasingly true with lower capital productivity α; a higher capital stock has a lower influence on production. The production to capital ratio ($\delta + g_R > g_K$) and consumption to capital ratio ($g_C > g_K$) are both increasing. The flow to stock ratio increases ($g_R > -V$). Capital accumulation is limited by the growth rate of consumption or the sum of the technical progress rate and the change rate in resource use.

If the economy starts with a relatively low capital and high use of resources then the upper optimal paths are considered. The production to capital ratio ($\delta + g_R < g_K$) and consumption to capital ratio ($g_C < g_K$) are both decreasing requiring a minimum growth rate of capital. The flow to stock ratio decreases ($g_R < -V$) implying a decline in resource use.

The Stiglitz model shows that technological progress is a solution to overcome the starvation of mankind. It does not, however, explain technological progress at all. The technological growth parameter δ falls like "manna from heaven". In reality technological progress is the result of innovation. Therefore, it should be treated as an endogenous variable. As previously mentioned an analysis of this endogenous growth theory applied to exhaustible resources exceeds the realm of this thesis. It seems unrealistic that a sustained level of consumption can be achieved with vanishingly small levels of the exhaustible resource input. As Krautkraemer (1998, p. 2093) pointed out, "a given amount of material output requires a minimum amount of material input, and unless material output goes to zero as the economy grows, some positive level of resource input must be maintained." A finite resource stock only allows the production of a finite output, leading again to the ultimate starvation of mankind. Only if the limited natural resource input is substituted by an infinite available resource as a result of technological progress, will the ultimate starvation of mankind be avoided.

5.2 Uncertain Technological Progress

A first attempt to model such an alternative resource-saving technological progress was done by Dasgupta and Heal (1974). They consider the possibility that a new technology will eventually appear that is not reliant on limited natural resources. This new technology is not the result of a smooth gradual innovation process, but rather a technological breakthrough arriving in a discrete once-for-all manner with economy-wide consequences. Specific examples are synthetic fuel or alternative energy sources like sunlight, wind, water, and geothermal heat. Dasgupta and Heal (1974, p. 19) suppose that "we know exactly the nature of the technical change that will occur, but we treat the date at which the event occurs as a random variable." In other words, a backstop technology is currently known, but its practical applicability still awaits a technological breakthrough. The unknown date T of the availability of the new technology is supposed to be random with an exogenous given probability density function ω_t:

$$prob(T = t) = \omega_t \tag{5.18}$$

$$\omega_t > 0 \tag{5.19}$$

$$\int_0^\infty \omega_t dt = 1. \tag{5.20}$$

The technological advance is costless and will certainly be discovered before the exhaustible resource is completely extracted. The new technology is characterized as being the discovery of an infinite commodity which is available at a constant rate N and utilized at rate Z_t. The available commodity (inventory Q_t) at each point in time is given by:

$$\dot{Q} = N - Z_t. \tag{5.21}$$

The economy goes through two different periods, before and after the technological breakthrough. After the technological breakthrough (for the period beyond T) the economy uses the new technology. The new commodity Z_t is a perfect substitute for the exhaustible resource R_t. The given inventory stock Q_T is thus given by:

$$Q_T = S_0 - \int_0^T R_t = S_T. \tag{5.22}$$

Additionally, the economy begins the second period with a given physical capital stock K_T. The depreciation rate of physical capital is negligible. Physical capital K_t and the infinite commodity Z_t are used to produce the output Y_t of the economy. The output is then either consumed or invested. Investment in physical capital increases its stock:

$$\dot{K} = P(K_t, Z_t) - C_t. \tag{5.23}$$

With enhanced production greater consumption in the future is possible. The Cobb-Douglas type production function is given by:

$$P(K_t, Z_t) = K^{\alpha} Z^{1-\alpha}. \tag{5.24}$$

The constraints that limit the choice of a consumption profile after the technological breakthrough are described by:

$$W(K_T, Q_T) = \max_{C,R} \int_T^{\infty} e^{-\rho(t-T)} U(C_t) dt \tag{5.25}$$

subject to the dynamic capital accumulation, the inventory development Q_t:

$$\dot{K} = P(K_t, Z_t) - C_t$$
$$\dot{Q} = N - Z_t$$

and the boundary and non-negativity constraints:

$$K(T) = K_T \ , \ K(t) \geq 0$$
$$Q(T) = Q_T \ , \ Q(t) \geq 0.$$

Based on the optimization problem, consumption and as a result welfare is maximized if the marginal product of capital P_K converges to ρ. The solution to the optimization problem is the optimal policy $Z_t = N$, i.e. the available commodity is utilized.

To answer the question how the economy behaves before the technological breakthrough, the complete optimization problem must be considered. The characterization of the economy before the technological progress is described in chapter 3. In short, a given stock of an exhaustible natural resource S_0 is continuously reduced by the flow R_t of resource extraction until the technological breakthrough is reached. In addition to the given natural resource stock, the economy starts with a given physical capital stock K_0. Physical capital K_t and the flow of resource extraction R_t are used to produce the output Y_t of the economy, which is then either consumed or invested. At time T, an additional utility stream $W(K_T, S_T)$ becomes available. In the words of Kamien and Schwartz (1978, p. 181) "the new technology need not be implemented immediately upon availability nor must exhaustible resource use diminish or cease upon employment of the new technology. It may be optimal to continue using the old one at a modified rate and gradually to employ the new one as it becomes economic." The additional utility stream could therefore depend on the stocks of capital and exhaustible resource remaining at T.

The expected present value of utility is maximized for optimal social welfare:

$$E \left[\max_{C,R} \int_0^\infty e^{-\rho t} U(C_t) dt \right] = \max_{C,R} \int_0^\infty e^{-\rho t} [\Omega_t U(C_t) + \omega_t W(K_t, S_t)] dt \quad (5.26)$$

subject to the dynamic capital accumulation, the process of resource extraction:

$$\dot{K} = F(K_t, R_t) - C_t$$
$$\dot{S} = -R_t$$

and the boundary conditions and non-negativity constraints:

$$K(0) = K_0 \, , \ K(t) \geq 0$$
$$S(0) = S_0 \, , \ S(t) \geq 0.$$

ω_t is the exogenous given probability that at time t the discovery date T of the new technology has been reached. Ω_t is the probability that at time t the discovery date T of the new technology has not been reached. The solution to the dynamic optimization problem uses the maximum principle of optimal control and combines the optimality conditions. The solution to this constrained optimization problem is derived in Appendix E. Note that the following solution is only true until the substitute is discovered. After the discovery date T the post technological breakthrough optimal policy is followed.

Using subscripts to denote partial derivatives and omitting the argument t, the first-order conditions are given by:

$$H_C := \psi_1 \overset{!}{=} e^{-\rho t} \Omega U'(C) \qquad (5.27)$$

$$H_K := -\dot{\psi}_1 \overset{!}{=} e^{-\rho t} \omega W_K + \psi_1 F_K \qquad (5.28)$$

$$H_R := \psi_2 \overset{!}{=} \psi_1 F_R \qquad (5.29)$$

$$H_S := -\dot{\psi}_2 \overset{!}{=} e^{-\rho t} \omega W_S. \qquad (5.30)$$

The optimal consumption and savings statement given by equation 5.27 is a slightly modified version of the basic Dasgupta-Heal model (see equation 3.5). The shadow price of capital equals the marginal utility if the technological breakthrough has not been reached. The relationship between the marginal opportunity cost of using one more unit of natural resource and employing it in final production equals the benefits (equation 5.29 is identical to 3.7). The marginal opportunity cost of using one more unit of physical capital (equation 5.28) is higher than that of the basic case (equation 3.6). This is due to the possible usability of physical capital for the new technology. In the basic Dasgupta-Heal model it is socially optimal to exhaust the natural resource (see equation 3.8). This may no longer be the case before the technological advancement. Equation 5.30 has been modified to include the possible usability of the natural resource stock after the technological breakthrough.

Combining the previously stated first-order conditions yields Ramsey's rule, which defines the optimal consumption path as:

$$\dot{C} = -[F_K - \rho]\frac{U'(C)}{U''(C)} - \frac{\omega}{\Omega}\frac{U'(C)}{U''(C)}\left[\frac{W_K}{U'(C)} - 1\right] \qquad (5.31)$$

$$\frac{\dot{C}}{C} = [\alpha x^{1-\alpha} - \rho] + \frac{\omega}{\Omega}[W_K C - 1]. \qquad (5.32)$$

ω_t/Ω_t is the conditional probability of the substitute being discovered at t given that it has not been discovered earlier. The optimal consumption path of the basic model (see equation 3.11) is slightly modified. The same holds true for the Hotelling rule:

$$\frac{\dot{F_R}}{F_R} = \frac{\omega}{\Omega} \frac{W_K}{U'(C)} + F_K - \frac{\omega}{\Omega} \frac{W_S}{U'(C)F_R} \tag{5.33}$$

$$\dot{x} = \frac{\omega}{\Omega} \frac{W_K}{U'(C)} \frac{1}{\alpha x^{-1}} + x^\alpha - \frac{\omega}{\Omega} \frac{W_S}{U'(C)} \frac{1}{\alpha(1-\alpha)x^{\alpha-1}}. \tag{5.34}$$

Since "the nature of the[se] path[s] (...) is far from obvious" (Dasgupta and Heal, 1974, p. 21), a specific example is analyzed.

Suppose the new technology is a highly efficient production method of synthetic fuel. Synthetic fuel can now be more efficiently produced than natural oil can be attained. In this example once the new technology has been discovered existing stocks of capital and exhaustible resources have no economic value. This assumption implies $W_K = W_S = 0$. The previously obtained results can be simplified to:

$$\frac{\dot{C}}{C} = \alpha x^{1-\alpha} - (\rho + \frac{\omega}{\Omega}) \tag{5.35}$$

$$\dot{x} = x^\alpha. \tag{5.36}$$

The basic equation for the ratio of capital to resource input is used (see chapter 3). The optimal consumption policy prior to the technological breakthrough pursues the solution described in the basic Dasgupta-Heal model, i.e. assumes the new technology will never be discovered.[27] The only difference is that the probability of the essential resource becoming inessential as a result of technological progress is added to the discount factor (time preference rate). This makes sense because the discount rate is higher in an uncertain situation, households will consume more today than in the future. In order to simplify comparison with the basic result, ω_t is supposed to be Poisson distributed:

$$\omega_t = \pi e^{-\pi t}. \tag{5.37}$$

---

[27] For a formal proof compare Appendix E.1.

The Poisson distribution makes sense, because Bernoulli trials (time T has been reached/ not reached) are considered over an infinite time. This leads to a constant conditional probability of the new technology arriving at t, given no previous occurrence of $\omega/\Omega = \pi$. The optimal consumption path is given by:

$$\frac{\dot{C}}{C} = \alpha x^{1-\alpha} - (\rho + \pi). \tag{5.38}$$

At point T there is no existing capital stock and no inventory of the exhaustible resource. The break with the past is complete. The optimal consumption paths for the described economy are depicted below.

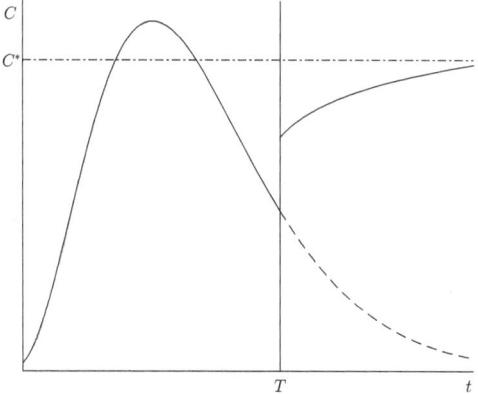

Fig. 5.3: Consumption Path Over Time in Case of a Backstop-Technology.
Source: Own illustration, based on Dasgupta and Heal (1974, p. 22).

Compared to the basic model, the first consumption path (until T) is similar. Consumption peaks earlier due to the higher time preference rate. After T the economy tends, in the long run, to the steady state consumption rate.

5.3 The Kamien-Schwartz Model

In the previous two models it is shown that technological progress along with increasing capital accumulation and substitution can prevent the starvation of mankind. Technological advancement proceeds steadily, without cost and exogenously in the Stiglitz model. In other words, it falls like "manna from heaven". Technological progress is still costless in the Dasgupta-Heal model but appears abruptly at an exogenous given random date.

In reality technological progress does not proceed smoothly and requires effort. A higher investment level increases the probability of a technological breakthrough but does not eliminate the randomness of innovation.

In order to account for a process of innovation, Kamien and Schwartz (1978) follow the framework used by Dasgupta and Heal (1974). A stylized characterization of the Kamien-Schwartz model is given below:

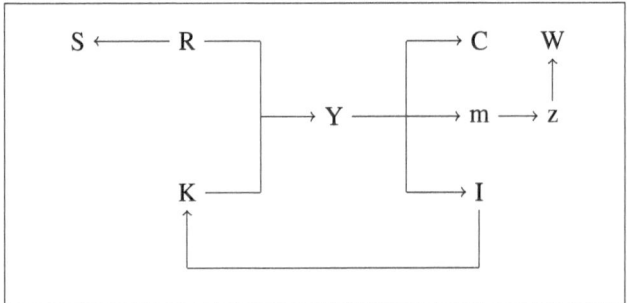

The economy undergoes two different periods, before and after the technological breakthrough. After T the economy uses the new technology. After the discovery date T the post technological breakthrough optimal policy is followed, described in the previous chapter.

The economy before the technological advancement is characterized by a given stock of an exhaustible natural resource S_0, which is continuously reduced by the flow R_t of resource extraction:

$$\dot{S} = -R_t. \tag{5.39}$$

In addition to the given natural resource stock, the economy starts with a given physical capital stock K_0. The depreciation rate of physical capital is negligible. The output Y_t of the economy is dependent on physical capital K_t and the natural resource extraction rate R_t. Output is either consumed or invested in physical capital or research. Consumption increases current well-being but leads to a negative change in capital. Investment in physical capital stock leads to enhanced production:

$$\dot{K} = F(K_t, R_t) - C_t - m_t. \tag{5.40}$$

With enhanced production greater consumption in the future is possible. The date of the technological breakthrough is unknown. However, it is affected by research effort, i.e. technological progress is made endogenous. Technological progress results from successful research and development (R&D) which requires resources to be diverted from consumption and capital investment into R&D at rate m_t. Once cumulative effective effort z_t in *R&D* is high enough, the new technology becomes available. An additional utility stream $W(K_T, S_T)$ becomes available from time T:

$$W(K_T, S_T) = \max_{C,R} \int_T^\infty e^{-\rho(t-T)} U(C_t) dt. \tag{5.41}$$

The discovery of an alternate more efficient production method makes existing stocks of capital and exhaustible resources irrelevant. This assumption implies $W(K_T, S_T) = W$, i.e. the additional utility stream W is independent of the previous capital and natural resource stock.

Effective effort is initially zero and accumulates with increasing *R&D* spending. Over time additional spending in *R&D* contributes less to the total effort (reflecting diminishing returns to *R&D* as *R&D* expenditure rise). Thus, the R&D rate is related to the growth of cumulative effective effort z_t devoted to the project by time t through a bounded, concave, monotone increasing and twice differentiable function $v(m_t)$ with the properties:

$$\dot{z} = v(m_t), \quad z(0) = 0 \tag{5.42}$$

$$v(0) = 0, \quad 0 < v'(m) < \infty, \quad v''(m) < 0. \tag{5.43}$$

The probability $\omega(z_t)$ that at time t the discovery date T of the new technology has been reached equals the probability that the R&D will be successfully completed by the time cumulative effort is z. Analogously to the Dasgupta-Heal model, the technological breakthrough is assumed to be certain, i.e. it will happen at some point in the future. $\Omega(z_t) = 1 - \omega(z_t)$ is the probability that at time t the discovery date T of the new technology has not been reached.

Further let h(z) be the conditional probability of completion:[28]

$$h(z) = \frac{\omega'(z)}{\Omega(z)} \tag{5.44}$$

$$h'(z) \geq 0 \quad for\ 0 \leq z \leq \bar{z}. \tag{5.45}$$

$\bar{z}$ is the smallest value of z for which completion is certain, i.e. $\omega(\bar{z}) = 1$. Furthermore, $\underline{z}$ is the minimum effort required for a possible technological breakthrough, i.e. $h(z) = \omega(z) = 0$ for efforts less than $\underline{z}$.

Households will receive utility U(C) as long as the old technology is in use. Considering a future utility stream with discounted value W at time t, the probability that the new technology will become available between two points in time (t,t+dt) is:

$$d\omega(z) = \omega'(z)\dot{z} = \omega'(z)v(m). \tag{5.46}$$

---

[28] Compare Appendix F.2 for a formal derivation.

The expected present value of utility is maximized for optimal social welfare:

$$E\left[\max_{C,R,m}\int_0^\infty e^{-\rho t}U(C_t)dt\right] = \max_{C,R,m}\int_0^\infty e^{-\rho t}[U(C_t)\{\Omega(z)\} + \omega'(z)v(m)W]dt$$

(5.47)

subject to the dynamic capital accumulation, the process of resource extraction, the process of innovation:

$$\dot{K} = F(K_t, R_t) - C_t - m_t$$
$$\dot{S} = -R_t$$
$$\dot{z} = v(m_t)$$

and the boundary conditions and non-negativity constraints:

$$K(0) = K_0 \ , \ K(t) \geq 0$$
$$S(0) = S_0 \ , \ S(t) \geq 0$$
$$z(0) = z_0 \ , \ z(t) \geq 0.$$

The solution to the dynamic optimization problem uses the maximum principle of optimal control and combines the optimality conditions. The solution to this constrained optimization problem is derived in Appendix F. Note that the following solution is only true until the substitute is discovered. At T the post technological breakthrough period of the economy begins.

Using subscripts to denote partial derivatives and omitting the argument t, the first-order conditions are given by:

$$H_C: \qquad e^{-\rho t} U'(C)\Omega(z) - \psi_1 \qquad \overset{!}{=} 0 \qquad\qquad (5.48)$$

$$H_K: \qquad\qquad \psi_1 F_K \qquad\qquad \overset{!}{=} -\dot{\psi}_1 \qquad\qquad (5.49)$$

$$H_R: \qquad\qquad \psi_1 F_R - \psi_2 \qquad \overset{!}{=} 0 \qquad\qquad (5.50)$$

$$H_S: \qquad\qquad\qquad 0 \qquad\qquad \overset{!}{=} -\dot{\psi}_2 \qquad\qquad (5.51)$$

$$H_m: \; e^{-\rho t}\omega'(z)v'(m)W - \psi_1 + \psi_3 v'(m) \; \overset{!}{\leq} 0 \qquad for\; m \geq 0 \quad (5.52)$$

$$H_z: -e^{-\rho t}U(C)\omega'(z) + e^{-\rho t}\omega''(z)v(m)W \overset{!}{=} -\dot{\psi}_3. \qquad (5.53)$$

Equation 5.52 needs further discussion. There are two possible solutions for m_t which maximize the Hamiltonian at each point in time. One policy is that of not undertaking R&D, i.e. $m = 0$. Then at $t = 0$:

(i) $\quad \psi_3(0)v'(0) - \psi_1(0) < 0,$

since $z = \omega'(z) = v(0) = 0$. There is no R&D when its marginal value is below the composite good's value in other uses. The other policy is to pursue R&D, i.e. $m > 0$. Then at $t = 0$:

(ii) $\quad \psi_3(0)v'(m) - \psi_1(0) = 0,$

which is the case while a technological breakthrough is being researched. To find the point where R&D begins, 5.53 is integrated, yielding:

$$\int_t^\infty \dot{\psi}_3 ds = \int_t^\infty e^{-\rho s}U(C)\omega'(z)ds - \int_t^\infty e^{-\rho s}\omega''(z)v(m)W ds \qquad (5.54)$$

$$\psi_3 = -e^{-\rho t}\omega'(z)W + \int_t^\infty e^{-\rho s}\omega'(z)[\rho W - U(C)]ds. \qquad (5.55)$$

Using this result, 5.52 can be written as:

$$v'(m)\int_t^\infty e^{-\rho s}\omega'(z)[\rho W - U(C)]ds \leq \psi_1. \qquad (5.56)$$

For $m > 0$ the marginal value of the composite good used in R&D must equal its marginal value in capital investment, equation 5.56. Inserting this solution into (i) gives:

(i) $v'(0) \int_0^\infty e^{-\rho s} \omega'(z)[\rho W - U(C)] ds < \psi_1(0).$

This inequality holds true if R&D is never undertaken, since then $\omega'(z) = 0$ and thus $\psi_3(0) = 0$. However, if R&D is undertaken at some point in the future the expected net utility from innovation is positive. The marginal value of cumulative effective R&D effort is greater than zero ($\psi_3(0) > 0$). There must be a value of ψ_1 at a point t_0 where the equality holds true, i.e. (ii) holds true.

To show this more easily (i) is rearranged to:

(i') $\psi_3(0)v'(0) - \frac{\psi_2(0)}{F_R} < 0.$

F_R increases over time, i.e. the value of the fraction decreases.[29] Hence, there will be a point t_0 where (i') becomes zero. At this point the marginal value of the composite good used in R&D equals the marginal value of the composite good in capital investment. The innovation process begins. Although R&D may not begin immediately at $t = 0$ it will begin in its imminent future. Due to the limited supply of exhaustible resources the pressure to immediately start R&D is high. For the case before the onset of R&D the result from chapter 3 can be applied. The case of undertaking R&D is analyzed in the following.

Equality (ii) must be true during R&D. For the case that households smooth consumption over time ($U(C) = ln(C)$) and using the Cobb-Douglas production functions $Y = F(K, R) = K^\alpha R^{1-\alpha}$, the optimal path of the capital to resource ratio is given by:

$$\dot{x} = x^\alpha. \tag{5.57}$$

---

[29] Note $\dot{\psi}_2 = \dot{\psi}_3 = 0$ for m=z=0.

This standard result needs no further discussion. In addition, the optimal consumption path is given by:

$$\frac{\dot{C}}{C} = \alpha x^{\alpha-1} - \rho - v(m)h(z).$$ (5.58)

The optimal effort in R&D is given by:

$$-\frac{v''(m)\dot{m}}{v'(m)} = \alpha x^{\alpha-1} - v'(m)h(z)C[\rho W - ln(C)].$$ (5.59)

Since $v'(m)$ is assumed to be positive and $v''(m)$ is assumed to be negative, the effort in R&D ($\dot{m}$) increases (decreases) when the right side of equation 5.59 is positive (negative). For example consider $v(m) = m^{0.5}$, then the left side reduces to $\frac{\dot{m}}{2m}$. The first part of the right side ($\alpha x^{\alpha-1}$) is decreasing over time due to the increasing substitution of capital for the exhaustible resource. The behavior of the second part is more complex. Consumption C and the effort in R&D (rate m) determines whether the second part is positive or negative.

The behavior of consumption is analyzed, equation 5.58. If the marginal product of capital is less than or equal to the time preference rate at time t_m (the time when R&D begins), then consumption decreases with increasing time. The cost function for R&D, v(m), and the conditional completion rate h(z) are both either zero[30] (for $m = 0$) or positive (for $m > 0$). If the marginal product of capital is larger than the time preference rate at time t_m, then consumption decreases (increases) as $m > m^0$ ($m < m^0$). m^0 is implicitly defined by equation 5.58. For a given value of x and z at each point in time, there is at most one value m^0 for which the growth rate of consumption is zero, i.e. consumption is maximized:

$$v(m^0(t)) = \frac{\alpha x^{\alpha-1} - \rho}{h(z)}.$$ (5.60)

---

[30] In this case the basic Dasgupta-Heal result is obtained.

At every point where the effort in $R\&D$ is higher than the critical value m^0, the growth rate of consumption is negative and vice versa. In the C-m space the value m^0 specifies the $\dot{C} = 0$ locus. Since $\alpha x^{\alpha-1}$ decreases over time and h(z) increases over time, the critical R&D rate m^0 decreases with time.[31] The $\dot{C} = 0$ locus moves down over time. Above each $\dot{C} = 0$ locus the effort in R&D is larger than m^0, leading to less consumption. More resources are diverted from consumption into R&D. Below each $\dot{C} = 0$ locus the effort in R&D is smaller than m^0, leading to higher consumption. The result of this partial analysis is depicted below.

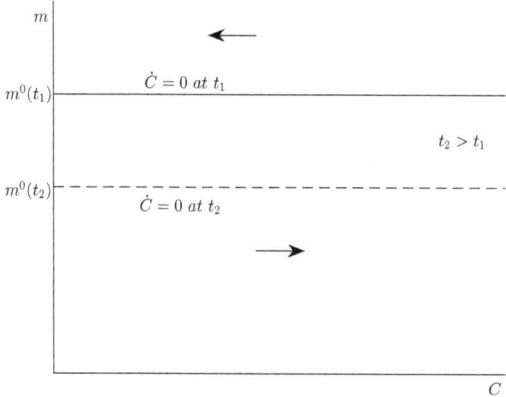

Fig. 5.4: Behavior of C.
Source: Own illustration, based on Kamien and Schwartz (1978, p. 186).

---

[31] Compare Appendix F.1 for a formal proof.

The behavior of the R&D rate is analyzed, equation 5.59. At date t_m the conditional probability of completion is $h(z) = 0$. Thus the effort in R&D increases over time until there is a sufficiently high probability of imminent completion. If this level is reached the effort is reduced. The maximum effort level is defined by $\dot{m} = 0$. At each point in time, for a given x and z, the combinations of C and m for which effort in R&D is maximized satisfy:[32]

$$\frac{\alpha x^{\alpha-1}}{v'(m)h(z)} = C[\rho W - ln(C)]. \tag{5.61}$$

The left side of equation 5.61 increases with m $\left(v'(m) = \frac{1}{2m^{0.5}}\right)$, while the right hand side is a concave function of C, i.e. G(C). The smallest value of the left side at any time is given by $\frac{x^{\alpha-1}}{v'(0)h(z)}$. G(C) is maximized at $C^* = e^{\rho W - 1}$. As long as $\frac{x^{\alpha-1}}{v'(0)h(z)} > C^*$ the effort in R&D increases. If this relationship is not true, then the effort in *R&D* behaves according to the shape of the $\dot{m} = 0$ locus. Equation 5.61 is considered an implicit definition of the function $m^C(C,t) = m$. The $\dot{m} = 0$ locus is described by positive growth rates for $0 < C < C^*$ and negative ones for $C^* < C$ at every point in time.[33] This result is applied to equation 5.59. If at one point in time the effort in *R&D* is higher than the critical value m^C (for which $\dot{m} = 0$), the growth rate of *R&D* effort is positive and vice versa. In the C-m space, for a given level of consumption the $\dot{m} = 0$ locus moves up over time.[34] The behavior of the R&D rate is plotted in the following graph.

---

[32] Under the assumption of $h(z) > 0$.
[33] Compare Appendix F.1 for a formal proof.
[34] Compare Appendix F.1 for a formal proof.

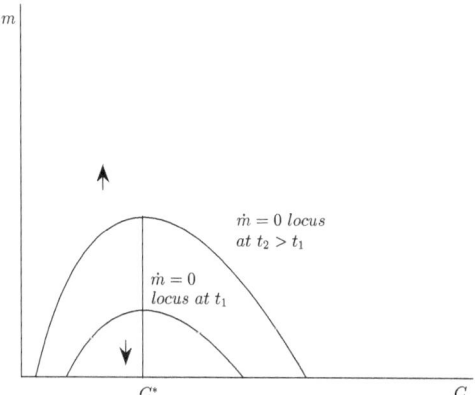

Fig. 5.5: Behavior of m.
 Source: Own illustration, based on Kamien and Schwartz (1978, p. 188).

There are three possible combinations for these obtained temporal patterns. In the first combination the marginal product of capital is less than or equal to the time preference rate at the date when research starts (t_m). If this point is at $t_m = 0$ households are focused on their well-being in the present, i.e. they consume more today and care less about tomorrow. Due to the limited supply of exhaustible resources consumption peaks and then declines continuously $(\dot{C} < 0)$. Consumption cannot be sustained without innovation. To hinder the starvation of mankind, effort in R&D is increased $(\dot{m} \geq 0)$ until there is a sufficiently high probability of imminent completion. If this level is reached effort is reduced. The resource stock is limited and continuously diminishing. The capacity of the economy is eventually reduced. If the technological breakthrough has not been achieved, this reduction in capacity leads to a decline in the consumption and R&D rate. Both functions are single peaked and decline towards zero. Consumption and R&D effort over time are depicted in the following figures for $t_m = 0$ and $t_m > 0$.

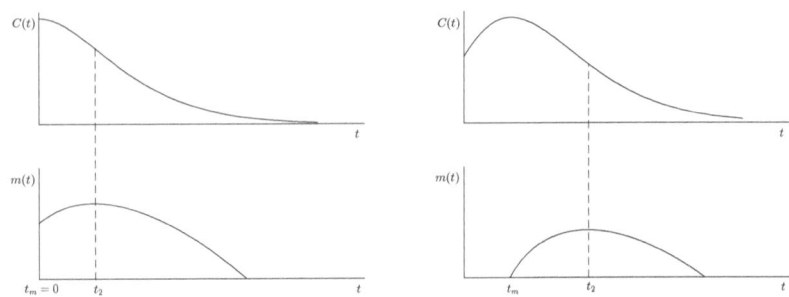

Fig. 5.6: Time Pattern for $F_K \leq rho$ at $t_m = 0$ and $t_m > 0$.
Source: Own illustration, based on Kamien and Schwartz (1978, p. 189).

For $t_m > 0$ consumption initially increases and then decreases. *R&D* effort is delayed until the consumption reaches its maximum. This level of consumption can no longer be sustained. *R&D* is undertaken to prevent the starvation of mankind. As the nonrenewable resource is depleted, the economy loses its capacity to invest in *R&D*. The projection of the optimal path in the C-m space is depicted in figure 5.7 for the case $t_m > 0$.

The optimal path is depicted by the line with arrows. Consumption falls over time (state (1)). After research is started effort in *R&D* is increased until either the technological breakthrough is reached or the capacity of the economy declines. The $\dot{m}$ locus moves up over time. "If the date of technological breakthrough T is not reached in the meantime, the $\dot{m}$ locus may overtake the optimal path. At the moment of intersection [state (2)], the optimal path is stationary $\dot{m} = 0$ while the $\dot{m}$ locus continues to rise" (Kamien and Schwartz, 1978, p. 188). Thereafter, the optimal path lies below the locus, consumption and effort in *R&D* decrease (state (3)).

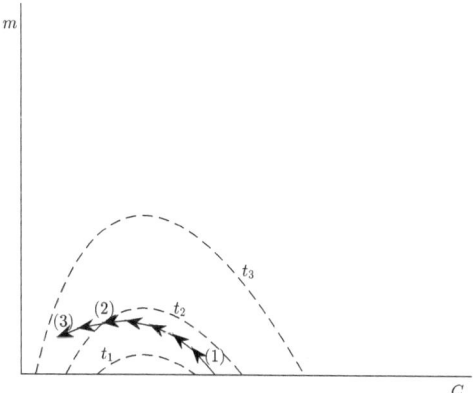

Fig. 5.7: Optimal Path for the Case of $F_K \leq rho$ at $t_m > 0$.
Source: Own illustration, based on Kamien and Schwartz (1978, p. 189).

In the second combination the marginal product of capital is larger than the time preference rate at the date when research starts (t_m). Households are focused on their future well-being, i.e. they save oil resources for their children. Consumption increases $(\dot{C} > 0)$ until it peaks. Effort in *R&D* increases until there is a sufficiently high probability of imminent completion. Once this level is reached effort is reduced. The resource stock is limited and diminishing, eventually reducing the capacity of the economy to invest in *R&D*. Both functions are single peaked and decline towards zero.

In the C-m space the $\dot{C}$ locus appears as the conditional probability of completion increases. The $\dot{C}$ locus moves down, while the $\dot{m}$ locus moves up over time. There are two possible cases. The optimal path can be overtaken first either by the falling $\dot{C}$ locus or the rising $\dot{m}$ locus. If consumption peaks before the *R&D* effort, the following behavior is observed.

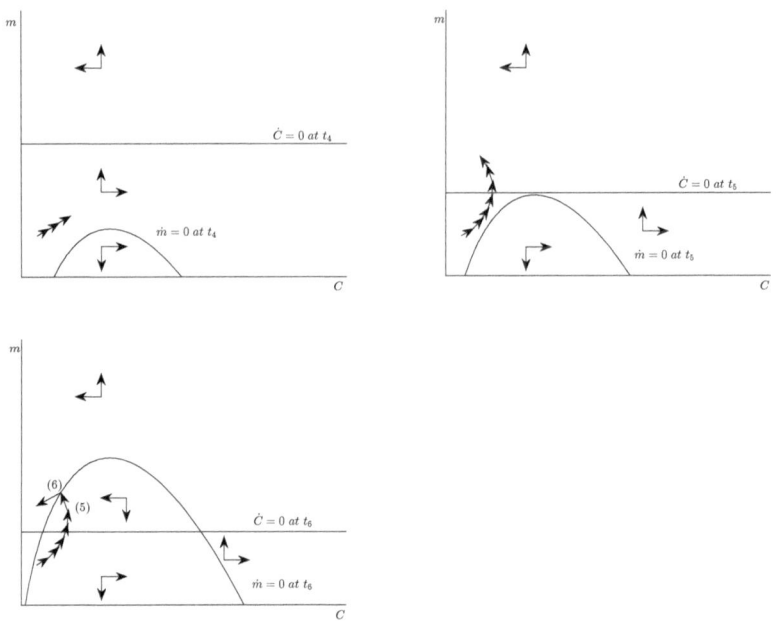

Fig. 5.8: Temporal Pattern for $F_K > \rho$ at t_m, When Consumption Peaks Before *R&D* Effort.
Source: Own illustration, based on Kamien and Schwartz (1978, p. 191).

Consumption increases until it crosses the falling $\dot{C}$ locus. Above this locus consumption decreases. If there is no technological breakthrough until this point, the optimal path also crosses the $\dot{m}$ locus. Both consumption and *R&D* effort decrease afterward.

The third combination supposes that the marginal product of capital is larger than the time preference rate at the date research starts (t_m) and the effort in *R&D* peaks first. Initially consumption and effort in *R&D* increase. The effort in *R&D* peaks and then declines. The consumption peaks at a later point after which both C and m fall over time until the technological breakthrough date is reached. The described behavior for this combination is depicted in figure 5.9.

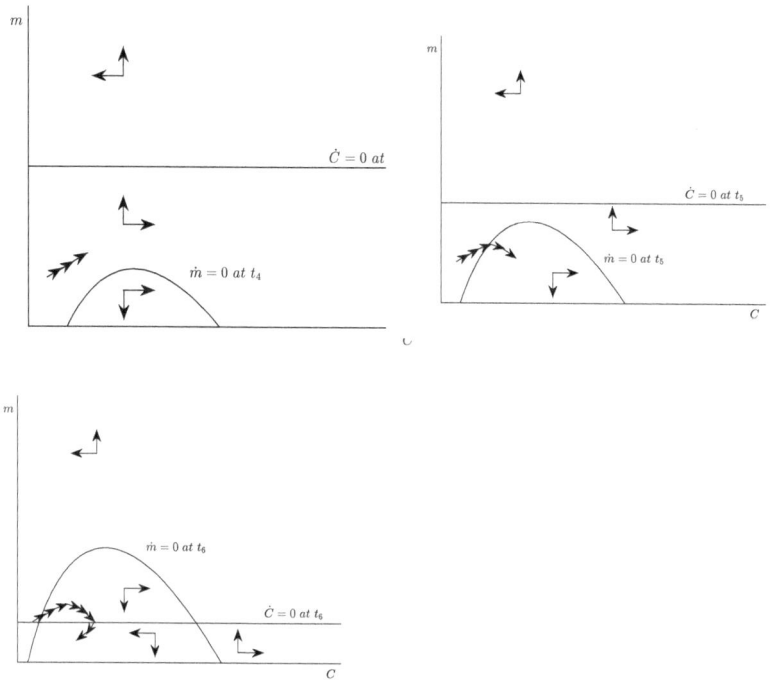

Fig. 5.9: Temporal Pattern for $F_K > \rho$ at t_m, When R&D Effort Peaks Before Consumption. Source: Own illustration, based on Kamien and Schwartz (1978, p. 191).

Both consumption and R&D are always single peaked. Consumption either falls continuously or grows initially and then declines depending on the marginal productivity of capital and the time preference rate ρ. R&D effort will fall towards zero as the nonrenewable resource is exhausted if a technological breakthrough has not been made. If the new technology is successfully implemented, the economy avoids any constraint. Once the new technology has been discovered the economy is totally new; no existing capital stock and no inventory of the exhaustible resource. The break with the past is complete.

6   Discussion and Conclusions

Today's society is extremely dependent on limited natural resources. The most obvious example is oil. In recent years, with the rise of the computer age, resources such as rare earth metals have begun to play an important role. As society grows more dependent little is known about their long-term effect on economic growth. This thesis presents the complex development that an economy based on limited resources will undergo.

In 1973, the world experienced an oil crisis, the price for oil increased from $3 per barrel to nearly $12. This crisis clearly demonstrated how dependent the global economy had become on a limited natural resource. In 1974, Dasgupta and Heal were one of the first who attempted to model economic development dependent on nonrenewable resources. Their model considered natural resources and physical capital as essential input factors for production. Due to a limited supply of nonrenewable resources this model resulted in the starvation of mankind. Despite being overly pessimistic, the Dasgupta-Heal model serves as a good basis for further theories.

As vital resources became more scarce and the production of waste increasingly problematic, recycling became relevant. In this thesis the expansion of the Dasgupta-Heal model by Pittel et al. (2010) to include recycling is detailed. In their model nature is no longer seen as part of the economy, but the economy is seen as part of nature. The material balance principle constrains economic production possibilities. For example, materials not used to create durable goods or recycled inputs must return to the environment as waste. This results in the need for the improved durability of products, capital accumulation, and perhaps most importantly the recovery and recycling of materials. Although recycling is theorized to extend the life of the economy, the ultimate result is still the starvation of mankind.

To avoid the starvation of mankind technological progress is needed. In this thesis the Dasgupta-Heal model is extended by an exogenously given resource-augmenting technological progress parameter. This is done using a modification of the theory by Stiglitz (1974). Resource-augmenting technological progress increases the efficiency of natural resource use, i.e. more output can be produced from a given amount of natural resources or less resources are needed for a given output. It seems, however, unrealistic that a sustained level of consumption can be achieved with vanishingly small levels of exhaustible resource input. A finite resource stock only allows the production of a finite output, leading again to the ultimate starvation of mankind. Only if the limited natural resource input is substituted by a backstop technology (an infinite available resource as a result of technological progress) will the ultimate starvation of mankind be avoided.

Dasgupta and Heal (1974) themselves extended their model to include such a backstop technology. Their model assumes that a technological breakthrough is inevitable but the exact date is unknown. In the period before the technological breakthrough, the optimal consumption policy assumes that the new technology will never be discovered. The only difference to the basic Dasgupta-Heal model is that the probability of the essential resource becoming inessential as a result of technological progress is added to the discount factor (time preference rate). After the technological breakthrough the stocks of the nonrenewable natural resource are worthless, the economy is considered "new".

In the Dasgupta-Heal model technological progress appears abruptly, sponta-neously, and without investment in *R&D* at an exogenous given random date. More realistically, technological breakthrough is the result of research and development. A higher investment level increases the probability of a technological breakthrough but does not eliminate the randomness of innovation. Kamien and Schwartz (1978) considered this more realistic case. They showed that both consumption and *R&D*

are always single peaked. Consumption either falls continuously or grows initially and then declines depending on the marginal productivity of capital and the time preference rate ρ. *R&D* effort will fall towards zero as the nonrenewable resource is exhausted if a technological breakthrough has not been made. If the new technology is successfully implemented, the economy avoids any constraint.

The next step is to consider recycling in a model with endogenous technological progress (backstop technology). A three stage model, which addresses an economy waiting for two backstop technologies, would be particularly interesting. The economy without recycling waits for recycling (first "backstop technology"), while expecting a superior backstop technology (alternative energy sources) in the long-run.

As proven by history economic growth and as a result mankind is dependent on technological progress. As society reaches a certain level of well-being, the acute need for research and development may be unclear. It is tempting to forgo investment in technological progress. In the example of oil, the incentive to do research on renewable energy is lower when the oil price is at e.g. $50 per barrel (high well-being) than when the oil price is high, e.g. $120 per barrel (lower well-being). As the model shows despite its current well-being, society must invest in *R&D* today to insure its continued well-being.

Appendix: A Sketch of Solutions

A   The Dasgupta-Heal Model

The present value of utility is maximized for optimal social welfare:

$$\max_{C,R} \int_0^\infty e^{-\rho t} U(C_t) dt \tag{A.1}$$

subject to the dynamic capital accumulation, the process of resource extraction:

$$\dot{K} = F(K_t, R_t) - C_t$$
$$\dot{S} = -R_t$$

and the boundary and non-negativity constraints:

$$K(0) = K_0 , \ K(t) \geq 0$$
$$S(0) = S_0 , \ S(t) \geq 0.$$

The present-value Hamiltonian in continuous time is defined as:

$$H = e^{-\rho t} U(C_t) + \psi_1 [F(K_t, R_t) - C_t] + \psi_2 [-R_t]. \tag{A.2}$$

ψ_1 denotes the shadow price of capital and ψ_2 the shadow price of the natural resource.

Using subscripts to denote partial derivatives and omitting the argument t, the first-order conditions are given by:

$$H_C : e^{-\rho t} U'(C) - \psi_1 \overset{!}{=} 0 \qquad\qquad (A.3)$$

$$H_K : \qquad \psi_1 F_K \qquad \overset{!}{=} -\dot{\psi}_1 \qquad\qquad (A.4)$$

$$H_R : \qquad \psi_1 F_R - \psi_2 \quad \overset{!}{=} 0 \qquad\qquad (A.5)$$

$$H_S : \qquad\quad 0 \qquad\quad \overset{!}{=} -\dot{\psi}_2. \qquad\qquad (A.6)$$

Differentiating A.3 with respect to time, dividing it by ψ_1 and combining the solution with equation A.4 yields the optimal consumption path (Ramsey rule):

$$\dot{C} = -[F_K - \rho] \frac{U'(C)}{U''(C)}. \qquad\qquad (A.7)$$

The constant inter-temporal elasticity of substitution (CIES) form of the utility function is given by:

$$U(C) = \begin{cases} \frac{C^{1-\theta}-1}{1-\theta} & \text{if } \theta > 0,\ \theta \neq 1 \\ ln(C) & \text{if } \theta = 1. \end{cases} \qquad\qquad (A.8)$$

For $\theta = 1$, the optimal consumption path is given by

$$\dot{C} = [F_K - \rho]C. \qquad\qquad (A.9)$$

The Hotelling rule can be derived by differentiating A.5 with respect to time, dividing it by ψ_2 and combining the solution with equations A.4 and A.6. This yields:

$$\dot{F}_R / F_R = F_K. \qquad\qquad (A.10)$$

If the production function is homogenous of degree one, it is possible to write $x \equiv K/R$ and $f(x) \equiv F(K/R, 1)$. x represents the capital to resource ratio. Using this definition, the elasticity of substitution between physical capital and the exhaustible resource is given by:

$$\sigma = -\left(\frac{dMRS}{dK/R}\right)^{-1}\frac{MRS}{K/R}$$

$$MRS = \frac{f'(x)}{f(x)-xf'x}$$

$$\frac{dMRS}{dx} = \frac{f''(x)f(x)}{[f(x)-xf'(x)]^2}$$

$$\sigma = -\frac{f'(x)[f(x)-xf'(x)]}{xf(x)f''(x)}. \tag{A.11}$$

Using $F_R = f(x) - xf'(x)$ and $F_K = f'(x)$, the Hotelling rule can be rearranged to:

$$\frac{\dot{x}f'(x) - \dot{x}f'(x) - x\dot{x}f''(x)}{f(x)-xf'(x)} = f'(x)$$

$$\dot{x} = -\frac{f'(x)[f(x)-xf'(x)]}{xf''(x)}\frac{xf(x)}{xf(x)}$$

$$\frac{\dot{x}}{x} = \sigma\frac{f(x)}{x}. \tag{A.12}$$

For the Cobb-Douglas case with $F = K^\alpha R^{1-\alpha} = (K/R)^\alpha R = x^\alpha R$ and $\sigma = 1$ Hotelling's rule can be reduced to the Bernoulli differential equation:

$$\alpha\dot{x}/x = \alpha x^{\alpha-1} \tag{A.13}$$

$$\dot{x} = x^\alpha. \tag{A.14}$$

Using the Bernoulli transformation to derive the linear ordinary differential equation (ODE) gives:

$$v(t) = x(t)^{1-\alpha} \tag{A.15}$$

$$\dot{v}(t) = 1 - \alpha. \tag{A.16}$$

Integration yields:

$$v(t) = v_0 + (1 - \alpha)t. \tag{A.17}$$

Transforming this back leads to the optimal time path of the capital to resource ratio as:

$$x_t^{1-\alpha} = [x_0^{1-\alpha} + (1 - \alpha)t]$$
$$x_t = [x_0^{1-\alpha} + (1 - \alpha)t]^{1/(1-\alpha)}. \tag{A.18}$$

In terms of the capital to resource ratio the optimal consumption path can be written as:

$$\dot{C} = [\alpha x_t^{\alpha-1} - \rho]C. \tag{A.19}$$

A.1 Starvation of Mankind

It can be shown that the growth rate of the capital-resource ratio is declining over time. Rewriting equation A.13 as:

$$\frac{\dot{x}}{x} = x^{\alpha-1}$$
$$\frac{\dot{x}}{x} = \left(x_0^{1-\alpha} + (1 - \alpha)t\right)^{\frac{\alpha-1}{1-\alpha}}. \tag{A.20}$$

Taking the derivative with respect to time, yields:

$$\frac{\partial \frac{\dot{x}}{x}}{\partial t} = \frac{\alpha - 1}{1 - \alpha}\left(x_0^{1-\alpha} + (1 - \alpha)t\right)^{\frac{\alpha-1}{1-\alpha}-1}(1 - \alpha)$$
$$\frac{\partial \frac{\dot{x}}{x}}{\partial t} = (\alpha - 1)\left(x_0^{1-\alpha} + (1 - \alpha)t\right)^{\frac{2(\alpha-1)}{1-\alpha}}. \tag{A.21}$$

Since α lies between 0 and 1, the first part is always lower than zero while the second part is always positive. The growth rate of the capital-resource ratio is declining over time.

It can be shown that consumption unambiguously converges to zero in the very long-run. Equation A.19 is rewritten as:

$$\frac{dC}{dt}/C = \alpha\frac{\dot{x}}{x} - \rho$$
$$\frac{dln(C)}{dt} = \alpha\frac{dln(x)}{dt} - \rho. \tag{A.22}$$

Integrating both sides from $t = 0$ to $t = t$ yields:

$$ln(C_t) - ln(C_0) = \alpha[ln(x_t) - ln(x_0)] - \rho t. \tag{A.23}$$

Applying the exponential function and substituting x_t leads:

$$C_t = C_0 x_0^{-\alpha}[x_0^{1-\alpha} + (1-\alpha)t]^{\frac{\alpha}{1-\alpha}}e^{-\rho t}. \tag{A.24}$$

To analyze the long-run behavior of consumption the limit is applied:

$$\lim_{t\to\infty} C_t = 0. \tag{A.25}$$

Consumption converges to zero in the very long-run.

A.2 Phase Diagram

To construct a phase diagram in the $(C/K, F/K)$ space the growth rates $g_{\frac{C}{K}}, g_{\frac{F}{K}}, g_C,$ and g_K are needed:

$$\frac{F}{K} = \left(\frac{K}{R}\right)^{\alpha-1}$$

$$g_{\frac{F}{K}} = \frac{\dot{F/K}}{F/K} = \frac{(\alpha-1)K^{\alpha-2}\dot{K}R^{1-\alpha} + (1-\alpha)K^{\alpha-1}R^{-\alpha}\dot{R}}{K^{\alpha-1}R^{1-\alpha}}$$

$$g_{\frac{F}{K}} = (\alpha-1)\frac{\dot{K}}{K} + (1-\alpha)\frac{\dot{R}}{R}$$

$$g_{\frac{F}{K}} = -(1-\alpha)(g_K - g_R). \tag{A.26}$$

With the side condition for capital accumulation from equation A.1, g_K is given by:

$$g_K = \frac{\dot{K}}{K} = \frac{F}{K} - \frac{C}{K}. \tag{A.27}$$

g_R can be obtained using equation A.10:

$$F_K = \dot{F}_R/F_R$$

$$\alpha\frac{F}{K} = \frac{(1-\alpha)\alpha K^{\alpha-1}\dot{K}R^{-\alpha} + (1-\alpha)(-\alpha)K^{\alpha}R^{-\alpha-1}\dot{R}}{(1-\alpha)K^{\alpha}R^{-\alpha}}$$

$$\alpha\frac{F}{K} = \alpha\frac{\dot{K}}{K} - \alpha\frac{\dot{R}}{R}$$

$$\frac{F}{K} = \frac{F}{K} - \frac{C}{K} - g_R$$

$$-\frac{C}{K} = g_R. \tag{A.28}$$

Using the penultimate results equation A.26 can be rewritten to:

$$g_{\frac{F}{K}} = -(1-\alpha)\frac{F}{K}. \tag{A.29}$$

$g_{\frac{C}{K}}$ is obtained using the side condition for capital accumulation from equation A.1:

$$\frac{C}{K} = \frac{F}{K} - \frac{\dot{K}}{K}$$

$$g_{\frac{C}{K}} = \frac{C\dot{/}K}{C/K} = \frac{\frac{\dot{C}}{K} - \frac{C}{K}\frac{\dot{K}}{K}}{C/K}$$

$$g_{\frac{C}{K}} = \frac{\dot{C}}{C} - \frac{F}{K} + \frac{C}{K}$$

$$g_{\frac{C}{K}} = -(1-\alpha)\frac{F}{K} + \frac{C}{K} - \rho. \tag{A.30}$$

In order to draw a phase diagram it is helpful to estimate the isoclines.

For $g_{\frac{C}{K}} = 0$, the following is true:

$$\frac{F}{K} = \frac{\frac{C}{K} - \rho}{1 - \alpha}, \tag{A.31}$$

and for $g_{\frac{F}{K}} = 0$, the following is true:

$$0 = -(1-\alpha)\frac{F}{K}. \tag{A.32}$$

In this case for all $\frac{C}{K}$, $\frac{F}{K} = 0$, i.e. the $g_{\frac{F}{K}}$ line equals the C/K axis.

For $g_C = 0$, the following is true:

$$\frac{F}{K} = \frac{\rho}{\alpha}, \tag{A.33}$$

i.e. the g_C line is a horizontal line.

For $g_K = 0$:

$$\frac{F}{K} = \frac{C}{K}, \tag{A.34}$$

i.e. the g_K line is a upward sloping linear line with slope 1.

The intersection point of the upward sloping $g_{\frac{C}{K}}$ line and the horizontal $g_{\frac{F}{K}}$ line is $(\rho, 0)$:

$$g_{\frac{C}{K}} = g_{\frac{F}{K}}$$
$$\frac{C}{K} = \rho.$$

(A.35)

B   A Closed-form Solution of the Dasgupta-Heal Model

A closed-form solution of the Dasgupta-Heal model can be derived by assuming an inter-temporal elasticity of substitution of $1/\theta = 1/\alpha$. Since there is no economic reason for this assumption and $\theta = 1 = \alpha$ would imply that only capital is used for production, the closed-form solution is derived for $1/\theta = 1/\alpha \neq 1$, i.e. $0 < \alpha < 1$. Hence, the following utility function is considered:

$$U(C) = \frac{C^{1-\theta} - 1}{1 - \theta}. \tag{B.1}$$

Similar to the case described in the previous section the optimal consumption path can be derived. For the above stated utility function equation A.19 only changes slightly to:

$$\frac{\dot{C}}{C} = \left[\frac{\alpha}{\theta} x^{\alpha-1} - \frac{\rho}{\theta} \right]. \tag{B.2}$$

Using A.13, the equation can be written as:

$$\frac{\dot{C}}{C} = \left[\frac{\alpha}{\theta} \frac{\dot{x}}{x} - \frac{\rho}{\theta} \right]. \tag{B.3}$$

To derive the closed-form solution B.3 is integrated from $t = 0$ to $t = t$, which yields:

$$ln(C_t) - ln(C_0) = \frac{\alpha}{\theta}[ln(x_t) - ln(x_0)] - \frac{\rho}{\theta}t$$

$$\frac{C_t}{C_0} = \left[\frac{x_t}{x_0}\right]^{\alpha/\theta} e^{-(\rho/\theta)t}. \tag{B.4}$$

Moreover, using the definition of x and equation A.13, gives:

$$\frac{\dot{x}}{x} = \frac{\dot{K}}{K} - \frac{\dot{R}}{R} = \frac{\dot{K}}{Rx} - \frac{\dot{R}x}{Rx}$$

$$x^\alpha R = \dot{K} - \dot{R}x$$

$$F(K,R) = \dot{K} - \dot{R}x. \tag{B.5}$$

Combining this result with the side condition for capital accumulation from equation A.1 yields:

$$C = -\dot{R}x$$

$$-\dot{R}_t = \frac{C_t}{x_t}. \tag{B.6}$$

Condition B.6 and B.4 with $\theta = \alpha$ give:

$$\frac{d^2S}{dt^2} = \frac{C_t}{x_t} = \frac{C_0}{x_0} e^{-(\rho/\alpha)t}. \tag{B.7}$$

Integrating equation B.7 twice yields:

$$S_t = \left[\frac{\alpha}{\rho}\right]^2 \frac{C_0}{x_0} e^{-(\rho/\alpha)t} + Bt + D, \tag{B.8}$$

where B and D are constants of integration. Since $lim_{t\to\infty}S_t = 0$, the constants of integration must be $B = D = 0$. Hence, the growth rate of the natural resource stock is described by:

$$\frac{\dot{S}}{S} = -\frac{\rho}{\alpha}. \tag{B.9}$$

The resource extraction process is given as a side condition in equation A.1. Using the previously obtained result the following holds true:

$$R_t = \frac{\rho}{\alpha}S_t = \frac{\rho}{\alpha}S_0 e^{-(\rho/\alpha)t} \tag{B.10}$$

$$\dot{R}_t = -\left(\frac{\rho}{\alpha}\right)^2 S_0 e^{-(\rho/\alpha)t}. \tag{B.11}$$

Making use of equation B.10 for $t = 0$ gives an alternative representation of the capital to resource ratio:

$$x(0) \equiv \frac{K_0}{R_0} = \frac{\alpha K_0}{\rho S_0}. \tag{B.12}$$

From A.18, B.6 and B.11 the optimal consumption path can be obtained as:

$$C_t = -\dot{R}x_t = \left(\frac{\rho}{\alpha}\right)^2 S_0 e^{-(\rho/\alpha)t} \left[\left(\frac{\alpha K_0}{\rho S_0}\right)^{1-\alpha} + (1-\alpha)t\right]^{1/(1-\alpha)} \tag{B.13}$$

$$ln(C_t) = ln(S_0) + 2\,ln\left(\frac{\rho}{\alpha}\right) + \frac{1}{1-\alpha}\,ln\left(\left(\frac{\alpha K_0}{\rho S_0}\right)^{1-\alpha} + (1-\alpha)t\right) - \frac{\rho}{\alpha}t. \tag{B.14}$$

Consumption $C(T)$ peaks when $dln(C_t)/dt = 0$:

$$\frac{dln(C)}{dt} = \frac{1-\alpha}{1-\alpha}\frac{1}{((\alpha K_0)/(\rho S_0))^{1-\alpha} + (1-\alpha)t} - \frac{\rho}{\alpha} \overset{!}{=} 0$$

$$T_C = \frac{(\alpha/\rho) - ((\alpha K_0)/(\rho S_0))^{1-\alpha}}{1-\alpha}, \tag{B.15}$$

where $T > 0$ if the ratio K_0/S_0 is sufficiently small.

Using the previously obtained results, the optimal time paths for capital and production can be derived:

$$
K_t = x_t R_t = \left(\frac{\rho}{\alpha}\right) S_0 e^{-(\rho/\alpha)t} \left[\left(\frac{\alpha K_0}{\rho S_0}\right)^{1-\alpha} + (1-\alpha)t\right]^{1/(1-\alpha)} \tag{B.16}
$$

$$
F_t = x_t^{\alpha} R_t = \left(\frac{\rho}{\alpha}\right) S_0 e^{-(\rho/\alpha)t} \left[\left(\frac{\alpha K_0}{\rho S_0}\right)^{1-\alpha} + (1-\alpha)t\right]^{\alpha/(1-\alpha)}. \tag{B.17}
$$

The stock of capital attains its peak at the same time as consumption whereas production attains its peak earlier, since $0 < \alpha < 1$:

$$
\frac{d\ln(F)}{dt} = \frac{\alpha(1-\alpha)}{1-\alpha} \frac{1}{((\alpha K_0)/(\rho S_0))^{1-\alpha} + (1-\alpha)t} - \frac{\rho}{\alpha} \overset{!}{=} 0
$$

$$
T_F = \frac{(\alpha^2/\rho) - ((\alpha K_0)/(\rho S_0))^{1-\alpha}}{1-\alpha} < T_C. \tag{B.18}
$$

C   Recycling Under a Material Balance Constraint

The present value of utility is maximized for optimal social welfare:

$$\max_{C,R,J} \int_0^\infty e^{-\rho t} U(C_t)\,dt \tag{C.1}$$

subject to the dynamic capital accumulation, the process of virgin resource extraction, the process of waste recycling:

$$\dot{K} = F(K_t, R_t, J_t) - C_t$$
$$\dot{S} = -R_t$$
$$\dot{D} = -J_t + (R_t + J_t)\kappa_t$$

and the boundary and non-negativity constraints:

$$K(0) = K_0 \ , \ K(t) \geq 0$$
$$S(0) = S_0 \ , \ S(t) \geq 0$$
$$D(0) = D_0 \ , \ D(t) \geq 0.$$

Note that $\kappa = \frac{C}{F(K,R,J)}$.

The present-value Hamiltonian in continuous time is defined as:

$$H = e^{-\rho t}U(C_t) + \psi_1\left[F(K_t, R_t, J_t) - C_t\right] + \psi_2\left[-R_t\right] + \psi_3\left[-J_t + (R_t + J_t)\kappa_t\right]. \quad \text{(C.2)}$$

Using subscripts to denote partial derivatives and omitting the argument t, the first-order conditions are given by:

$$H_C: \qquad e^{-\rho t}U'(C) - \psi_1 + \psi_3\mu \quad \overset{!}{=} 0 \qquad\qquad \text{(C.3)}$$

$$H_K: \qquad\quad \psi_1 F_K - \psi_3\mu\kappa F_K \quad \overset{!}{=} -\dot{\psi}_1 \qquad\qquad \text{(C.4)}$$

$$H_R: \psi_1 F_R - \psi_2 + \psi_3\kappa - \psi_3\mu\kappa F_R \overset{!}{=} 0 \qquad\qquad \text{(C.5)}$$

$$H_S: \qquad\qquad\qquad 0 \qquad\quad \overset{!}{=} -\dot{\psi}_2 \qquad\qquad \text{(C.6)}$$

$$H_J: \psi_1 F_J - \psi_3 + \psi_3\kappa - \psi_3\mu\kappa F_J \overset{!}{=} 0 \qquad\qquad \text{(C.7)}$$

$$H_D: \qquad\qquad\qquad 0 \qquad\quad \overset{!}{=} -\dot{\psi}_3, \qquad\qquad \text{(C.8)}$$

where $\mu = \frac{R+J}{F(K,R,J)}$.

For simplicity purposes the composite variable $v = e^{-\rho t}U'(C)$ is introduced. Using the composite variable and differentiating C.3 with respect to time and combining the solution with equations C.4 and C.8 yields:

$$\dot{v} + \dot{\psi}_3\mu + \psi_3\dot{\mu} = -\psi_1 F_K + \psi_3\mu\kappa F_K$$

$$\dot{v} = -(v + \psi_3\mu)F_K + \psi_3\mu\kappa F_K - \psi_3\dot{\mu}$$

$$\frac{\dot{v}}{v} = -F_K - \frac{\psi_3}{v}\left[F_K\mu(1-\kappa) + \dot{\mu}\right], \qquad\qquad \text{(C.9)}$$

where the shadow price ratio $\frac{\psi_3}{v}$ is obtained by inserting $\psi_1 = v + \psi_3\mu$ into equation C.7 and solving for the ratio. For the logarithmic utility $(U(C) = ln(C))$ and the production function $F(K, R, J) = K^\alpha R^\beta J^\phi$, the optimal consumption path

(Ramsey's rule) is given by:

$$\frac{\dot{C}}{C} = F_K - \rho + \frac{F_J \mu}{(1-\kappa)(1-F_J\mu)}\left[F_K(1-\kappa) + \frac{\dot{\mu}}{\mu}\right] \qquad (C.10)$$

$$\frac{\dot{C}}{C} = \alpha\frac{F}{K} - \rho + \frac{\phi\frac{F}{J}\mu}{(1-\kappa)(1-\phi\frac{F}{J}\mu)}\left[\alpha\frac{F}{K}(1-\kappa) + \frac{\dot{\mu}}{\mu}\right]. \qquad (C.11)$$

In order to derive the Hotelling rule for recycled waste, equation C.7 is solved for F_J:

$$F_J = \frac{1-\kappa}{\psi_1\psi_3^{-1} - \mu\kappa}.$$

Differentiating this expression with respect to time and with equations C.4 and C.8 gives:

$$\frac{\dot{F}_J}{F_J} = -\frac{\dot{\kappa}}{1-\kappa} - \left(\frac{\dot{\psi}_1\psi_3^{-1} - (\dot{\mu}\kappa + \dot{\kappa}\mu)}{\psi_1\psi_3^{-1} - \mu\kappa}\right).$$

Using $\frac{1}{\psi_1\psi_3^{-1}-\mu\kappa} = \frac{F_J}{1-\kappa}$ from equation C.7 and substituting $\dot{\psi}_1$ (equation C.4), this simplifies to:

$$\frac{\dot{F}_J}{F_J} = -\frac{\dot{\kappa}}{1-\kappa} - \frac{F_J}{1-\kappa}\left[-\frac{\psi_1}{\psi_3}F_K + \mu\kappa F_K - (\dot{\mu}\kappa + \dot{\kappa}\mu)\right]. \qquad (C.12)$$

The fraction $\frac{\psi_1}{\psi_3}$ is obtained from equation C.7 and equals $\frac{1-\kappa+\mu\kappa F_J}{F_J}$. The Hotelling rule for recycled waste is given by:

$$\frac{\dot{F}_J}{F_J} = F_K + \left[\frac{F_J}{1-\kappa}(\dot{\mu}\kappa + \dot{\kappa}\mu) - \frac{\dot{\kappa}}{1-\kappa}\right]. \qquad (C.13)$$

The Hotelling rule for virgin resources is obtained similar. Solving equation C.5 for F_R and adding "zero", i.e. adding and subtracting ψ_3, yields:

$$F_R = \frac{\psi_2 - \psi_3 + \psi_3(1-\kappa)}{\psi_1 - \mu\kappa\psi_3}. \tag{C.14}$$

Differentiating this expression with respect to time and making use of equation C.4, C.6 and C.8 gives:

$$\frac{\dot{F}_R}{F_R} = \frac{-\psi_3\dot{\kappa}}{\psi_2 - \psi_3 + \psi_3(1-\kappa)} - \frac{\dot{\psi}_1 - \psi_3(\dot{\mu}\kappa + \dot{\kappa}\mu)}{\psi_1 - \psi_3\mu\kappa}. \tag{C.15}$$

Making use of equation C.14, yields:

$$\frac{\dot{F}_R}{F_R} = \frac{1}{\psi_1\psi_3^{-1} - \kappa\mu}\left[-\frac{\dot{\kappa}}{F_R} + (\psi_1\psi_3^{-1} - \mu\kappa)F_K + (\dot{\mu}\kappa + \dot{\kappa}\mu)\right]. \tag{C.16}$$

Since $\frac{1}{\psi_1\psi_3^{-1} - \kappa\mu} = \frac{F_J}{1-\kappa}$, Hotelling's rule for virgin resources is given by:

$$\frac{\dot{F}_R}{F_R} = F_K + \left[\frac{F_J}{1-\kappa}(\dot{\mu}\kappa + \dot{\kappa}\mu) - \frac{\dot{\kappa}}{1-\kappa}\frac{F_J}{F_R}\right]. \tag{C.17}$$

C.1 Long-run Equilibrium of the Economy

To derive the long-run growth rates for consumption and output, equation C.7 is rearranged to:

$$\psi_3 = \frac{\psi_1 F_J}{1 - \kappa + \mu\kappa F_J}$$

$$\psi_3 = \frac{\psi_1 F\phi}{J(1-\kappa) + \phi(R+J)\kappa}. \tag{C.18}$$

Inserting this result in C.3 and simplifying, yields:

$$-e^{-\rho t}U'(C) = \psi_1 \left[\frac{-1+\phi(R+J)}{J(1-\kappa)+\phi(R+J)\kappa} \right] \equiv \psi_1 B. \tag{C.19}$$

Differentiating this expression with respect to time and dividing by itself gives:

$$\rho + \frac{\dot{C}}{C} = \frac{\dot{\psi_1}}{\psi_1} + \frac{\dot{B}}{B}. \tag{C.20}$$

Since in steady state $\dot{C} = \dot{F} = \dot{\kappa} = 0$ equation C.13 is identical to equation C.17. Thus $g_R = g_J$ in steady state:

$$\frac{\dot{F_R}}{F_R} = \frac{\dot{F_J}}{F_J} \tag{C.21}$$

$$\alpha\frac{\dot{K}}{K} + (\beta-1)\frac{\dot{R}}{R} + \phi\frac{\dot{J}}{J} + \frac{F_J}{1-\kappa} = \alpha\frac{\dot{K}}{K} + \beta\frac{\dot{R}}{R} + (\phi-1)\frac{\dot{J}}{J} + \frac{F_J}{1-\kappa} \tag{C.22}$$

$$\frac{\dot{R}}{R} = \frac{\dot{J}}{J}. \tag{C.23}$$

Using this penultimate result, it can be shown that in steady state $g_B = 0$. Equation C.20 simplifies to:

$$g_C + \rho = g_{\psi_1}. \tag{C.24}$$

Moreover, rearranging C.5 gives:

$$\psi_2 - \psi_3\kappa = F_R(\psi_1 - \psi_3\mu\kappa) \tag{C.25}$$

$$\psi_2 - \psi_3 + \psi_3(1-\kappa) = F_R(\psi_1 - \psi_3\mu\kappa) \tag{C.26}$$

$$\psi_2 - \psi_3 = F_R\psi_1 - \psi_3[F_R\mu\kappa + (1-\kappa)]. \tag{C.27}$$

Using equation C.18, this is equivalent to:

$$\psi_2 - \psi_3 = \psi_1 \beta \frac{F}{R} - \psi_1 \frac{\phi F}{J(1-\kappa) + \phi \kappa (R+J)} [\beta \frac{F}{R} \mu \kappa + (1-\kappa)] \quad (C.28)$$

$$\psi_2 - \psi_3 = \psi_1 \beta \frac{F}{R} \left[1 - \frac{\phi}{\beta} \left[\frac{\beta(R+J)\kappa + R(1-\kappa)}{J(1-\kappa) + \phi(R+J)\kappa} \right] \right]. \quad (C.29)$$

In the steady state κ is constant and $g_R = g_J$, so the term in brackets is constant:

$$g_{\psi_1} = g_R - g_Y. \quad (C.30)$$

Deriving the growth rates from the production function gives:

$$g_Y = \alpha g_K + \beta g_R + \phi g_J. \quad (C.31)$$

Since in steady state $g_Y = g_K$ and $g_R = g_J$ this simplifies to:

$$(\beta + \phi) g_R = (1 - \alpha) g_Y. \quad (C.32)$$

$$g_R = g_Y \quad (C.33)$$

Combining equations C.24, C.30 and C.33 yields the growth rate of output in the long-run:

$$g_C^* = -\rho = g_Y^*. \quad (C.34)$$

The economy will not survive.

D   Exogenous Technological Progress

The considered production function includes only resource-augmenting technological progress:

$$F(K,R) = K^\alpha (AR)^{1-\alpha} , \qquad (\text{D.1})$$

where the technological level $A = A_0 e^{\delta t}$ is growing at the given constant and exogenous rate δ. In addition, a logarithmic utility function is assumed. For the purpose of simplicity, capital and output are considered in efficiency units, i.e. $k = K/A$ and $y = Y/A = \tilde{F}(k,R)$. The side condition of capital accumulation is modified to:

$$\dot{K} = F(K,AR) - C$$
$$\left[\frac{K}{A}\right]^{\cdot} = \dot{k} = \frac{\dot{K}}{A} - \frac{\dot{A}}{A}\frac{K}{A}$$
$$\dot{k} = \frac{F(K,AR)}{A} - \frac{C}{A} - \delta k$$
$$\dot{k} = \tilde{F}(k,R) - Ce^{-\delta t} - \delta k$$
$$\dot{k} = k^\alpha R^{1-\alpha} - c - \delta k. \qquad (\text{D.2})$$

Consumption in efficiency units is given by $c = \frac{C}{A} = Ce^{-\delta t}$.

The present value of utility is maximized for optimal social welfare:

$$\max_{C,R} \int_0^\infty e^{-\rho t} U(C_t) dt \tag{D.3}$$

subject to the dynamic capital accumulation, the process of resource extraction:

$$\dot{k} = \tilde{F}(k_t, R_t) - C_t e^{-\delta t} - \delta k_t$$
$$\dot{S} = -R_t$$

and the boundary and non-negativity constraints:

$$k(0) = k_0 \, , \; k(t) \geq 0$$
$$S(0) = S_0 \, , \; S(t) \geq 0.$$

The first-order conditions are given by:

$$H_C: \quad e^{-\rho t} U'(C) - \psi_1 e^{-\delta t} \overset{!}{=} 0 \tag{D.4}$$

$$H_k: \quad \psi_1 \alpha k^{\alpha-1} (R)^{1-\alpha} - \delta \overset{!}{=} -\dot{\psi}_1 \tag{D.5}$$

$$H_R: \psi_1 (1-\alpha) k^\alpha (R)^{-\alpha} - \psi_2 \overset{!}{=} 0 \tag{D.6}$$

$$H_S: \qquad\qquad 0 \qquad\quad \overset{!}{=} -\dot{\psi}_2. \tag{D.7}$$

Differentiating D.4 with respect to time, dividing it by ψ_1 and combining the solution with equation D.5 yields the optimal consumption path (Ramsey's rule):

$$\frac{\dot{C}}{C} = [\alpha k^{\alpha-1} R^{1-\alpha} - \rho]. \tag{D.8}$$

In efficiency units the optimal consumption path is given by:

$$\frac{\dot{c}}{c} = [\alpha k^{\alpha-1} R^{1-\alpha} - \rho - \delta]. \tag{D.9}$$

Consumption is feasible in the long-run if $\alpha k^{\alpha-1} R^{1-\alpha} = \tilde{F}_k = \rho$. As a result, if t approaches infinity $d\tilde{F}_k/dt$ must be zero. This yields:

$$\lim_{t \to \infty} \frac{\dot{\tilde{F}}_k}{\tilde{F}_k} = -(1-\alpha)\frac{\dot{k}}{k} + (1-\alpha)\frac{\dot{R}}{R} \overset{!}{=} 0$$

$$\frac{\dot{R}}{R} = \frac{\dot{k}}{k}$$

$$\frac{\dot{R}}{R} = \frac{\dot{K}}{K} - \frac{\dot{A}}{A}. \tag{D.10}$$

Further, the modified Hotelling rule can be derived by differentiating equation D.6 with respect to time and combining this derivative with equations D.5 and D.7:

$$\frac{\dot{\tilde{F}}_R}{\tilde{F}_R} = \tilde{F}_k - \delta. \tag{D.11}$$

In steady state $\tilde{F}_k = \delta$ holds true. In combination with Ramsey's rule consumption is only feasible in the long-run if $\delta \geq \rho$:

$$\frac{\dot{C}}{C} = \delta - \rho. \tag{D.12}$$

D.1 Phase Diagram

To derive the steady state solution it proves useful to set $\tilde{k} = K/AR$ and $\tilde{y} = F/AR = f(\tilde{k})$, i.e. the capital stock and output in efficiency units adjusted by the resource input are considered. The following relationship holds true:

$$\tilde{F}_k = \alpha \left[\frac{k}{R}\right]^{\alpha-1} = \alpha \left[\frac{K}{AR}\right]^{\alpha-1} = \alpha \tilde{k}^{\alpha-1} = f_{\tilde{k}}$$

$$\tilde{F}_R = (1-\alpha)\left[\frac{k}{R}\right]^{\alpha} = (1-\alpha)\tilde{k}^{\alpha} = f(\tilde{k}) - \alpha f(\tilde{k}). \tag{D.13}$$

$\dot{\tilde{k}}$ is given by:

$$\dot{\tilde{k}} = \left[\frac{K}{AR}\right]^{\cdot} = \frac{\dot{K}}{AR} - \frac{\dot{A}}{A}\tilde{k} - \frac{\dot{R}}{R}\tilde{k}$$

$$\dot{\tilde{k}} = f(\tilde{k}) - \frac{C}{AR} - \delta\tilde{k} - \frac{\dot{R}}{R}\tilde{k}. \tag{D.14}$$

The growth rate of $\tilde{k}$ is thus given by:

$$\frac{\dot{\tilde{k}}}{\tilde{k}} = \frac{f(\tilde{k})}{\tilde{k}} - \frac{\tilde{c}}{\tilde{k}} - \delta - \frac{\dot{R}}{R}. \tag{D.15}$$

$\dot{\tilde{c}}$ is given by:

$$\dot{\tilde{c}} = \left[\frac{C}{AR}\right]^{\cdot} = \frac{\dot{C}}{AR} - \delta\frac{C}{AR} - \frac{\dot{R}}{R}\frac{C}{AR}.$$

The growth rate of $\tilde{c}$ is given by:

$$\frac{\dot{\tilde{c}}}{\tilde{c}} = \frac{\dot{C}}{C} - \delta - \frac{\dot{R}}{R}. \tag{D.16}$$

Using Ramsey's rule, equation D.8, $g_{\tilde{c}}$ can be rewritten as:

$$g_{\tilde{c}} = \frac{\dot{\tilde{c}}}{\tilde{c}} = f_{\tilde{k}} - \rho - \delta - \frac{\dot{R}}{R}. \tag{D.17}$$

From Hotelling's rule, equation D.11, the growth rate of R can be derived:

$$\alpha\frac{\dot{R}}{R} = -\alpha\left[\frac{K}{AR}\right]^{\alpha-1} + \delta + \alpha\frac{F(K,AR)}{K} - \alpha\frac{C}{K} - \alpha\delta$$

$$\frac{\dot{R}}{R} = -\tilde{k}^{\alpha-1} + \frac{\delta}{\alpha} + \tilde{k}^{\alpha-1} - \frac{C}{K} - \delta$$

$$\frac{\dot{R}}{R} = \frac{1-\alpha}{\alpha}\delta - \frac{C}{K}$$

$$\frac{\dot{R}}{R} = \frac{1-\alpha}{\alpha}\delta - \frac{\tilde{c}}{\tilde{k}}. \tag{D.18}$$

Using equation D.18, the growth rate of $\tilde{k}$ simplifies to:

$$\frac{\dot{\tilde{k}}}{\tilde{k}} = \frac{f(\tilde{k})}{\tilde{k}} - \frac{\delta}{\alpha} \tag{D.19}$$

$$\frac{\dot{\tilde{k}}}{\tilde{k}} = \frac{f_{\tilde{k}} - \delta}{\alpha}. \tag{D.20}$$

The growth rate of $\tilde{c}$ simplifies to:

$$\frac{\dot{\tilde{c}}}{\tilde{c}} = f_{\tilde{k}} - \rho - \frac{\delta}{\alpha} + \frac{\tilde{c}}{\tilde{k}}. \tag{D.21}$$

In order to construct a phase diagram in the $(C/K, F/K)$ space the growth rates $g_{\frac{C}{K}}, g_{\frac{F}{K}}, g_C$, and g_K are needed. $g_{\frac{F}{K}}$ is given by:

$$g_{\frac{f}{k}} = g_{\frac{F}{K}} = \frac{\dot{f}}{f} - \frac{\dot{\tilde{k}}}{\tilde{k}}$$

$$g_{\frac{F}{K}} = (\alpha - 1)\frac{\dot{\tilde{k}}}{\tilde{k}}$$

$$g_{\frac{F}{K}} = -(1 - \alpha)\frac{f_{\tilde{k}} - \delta}{\alpha}$$

$$g_{\frac{F}{K}} = (1 - \alpha)\frac{\delta - f_{\tilde{k}}}{\alpha}$$

$$g_{\frac{F}{K}} = -(1 - \alpha)\frac{F}{K} + \frac{1 - \alpha}{\alpha}\delta. \tag{D.22}$$

$g_{\frac{C}{K}}$ is given by:

$$g_{\frac{\tilde{c}}{\tilde{k}}} = g_{\frac{C}{K}} = \frac{\dot{\tilde{c}}}{\tilde{c}} - \frac{\dot{\tilde{k}}}{\tilde{k}}$$

$$g_{\frac{C}{K}} = f_{\tilde{k}} - \rho - \frac{\delta}{\alpha} + \frac{\tilde{c}}{\tilde{k}} - \frac{f_{\tilde{k}}}{\alpha} + \frac{\delta}{\alpha}$$

$$g_{\frac{C}{K}} = -(1 - \alpha)\frac{F}{K} - \rho + \frac{C}{K}. \tag{D.23}$$

Using the preliminary result, the isoclines can be estimated. $g_{\frac{C}{K}} = 0$ yields:

$$\frac{F}{K} = \frac{\frac{C}{K} - \rho}{1 - \alpha}.$$ (D.24)

$g_{\frac{F}{K}} = 0$ yields:

$$\frac{F}{K} = \frac{\delta}{\alpha}.$$ (D.25)

The $g_{\frac{F}{K}}$ line is a horizontal line at $\frac{\delta}{\alpha}$.
  $g_C = 0$ yields:

$$\frac{F}{K} = \frac{\rho}{\alpha}.$$ (D.26)

The g_C line is a horizontal line.
  $g_K = 0$ yields:

$$\frac{F}{K} = \frac{C}{K}.$$ (D.27)

The g_K line is a upward sloping linear line with slope 1.
  The intersection point of the $g_{\frac{C}{K}}$ line and the $g_{\frac{F}{K}}$ line is $(\delta\frac{1-\alpha}{\alpha} + \rho, \frac{\delta}{\alpha})$:

$$g_{\frac{C}{K}} = g_{\frac{F}{K}}$$
$$\frac{C}{K} = \delta\frac{1-\alpha}{\alpha} + \rho.$$ (D.28)

In order to construct a phase diagram in the $(C/K, V)$ space the growth rate g_V is needed, where $V = R/S$, the ratio of resource utilization to the stock of the resource:

$$g_V = g_R - \frac{\dot{S}}{S}$$

$$g_V = g_R - \frac{(-R)}{S}$$

$$g_V = \delta\frac{1-\alpha}{\alpha} - \frac{C}{K} + V. \tag{D.29}$$

$g_V = 0$ yields:

$$\frac{C}{K} = \delta\frac{1-\alpha}{\alpha} + V. \tag{D.30}$$

$g_{\frac{F}{K}} = 0$ yields:

$$\frac{F}{K} = \frac{\delta}{\alpha}. \tag{D.31}$$

The $g_{\frac{C}{K}}$ line is a vertical line in this phase diagram.

Using $g_{\frac{F}{K}} = 0$, the intersection point of the $g_{\frac{C}{K}}$ line and the g_V line is given by $(\delta\frac{1-\alpha}{\alpha} + \rho, \rho)$:

$$g_{\frac{C}{K}} = g_{\frac{F}{K}}$$

$$\frac{C}{K} = \delta\frac{1-\alpha}{\alpha} + \rho$$

$$g_{\frac{C}{K}} = g_V$$

$$\delta\frac{1-\alpha}{\alpha} + \rho = \delta\frac{1-\alpha}{\alpha} + V$$

$$V = \rho. \tag{D.32}$$

E   Uncertain Technological Progress

The dynastic household faces the following optimization problem, which is
solved recursively. First, the optimization problem is solved for the time after
the innovation:

$$W(K_T, Q_T) = \max_{C,R} \int_T^\infty e^{-\rho(t-T)} U(C_t) dt \qquad \text{(E.1)}$$

subject to the dynamic capital accumulation, the inventory development Q_t:

$$\dot{K} = P(K_t, Z_t) - C_t$$
$$\dot{Q} = N - Z_t$$

and the boundary and non-negativity constraints:

$$K(T) = K_T \; , \; K(t) \geq 0$$
$$Q(T) = Q_T \; , \; Q(t) \geq 0.$$

Using the Hamiltonian, the solution is straight-forward, yielding:

$$\dot{C} = -[P_K - \rho] \frac{U'(C)}{U''(C)} \qquad \text{(E.2)}$$
$$\dot{Q} = 0. \qquad \text{(E.3)}$$

Consumption and thus welfare is maximized for $P_K = \rho$. In addition, $Z_t = N$ for all
$t \geq T$. Since the flow of the durable commodity is constant and exogenously given
by N, the production function can be restated as $P(K, N) = p(K)$. The economy

converges towards a steady state defined by:

$$p'(K^*) = \rho \qquad (\text{E.4})$$

$$C^* = p(K^*). \qquad (\text{E.5})$$

Second, using this preliminary result the complete optimization problem is stated and solved as follows. The expected present value of utility is given by:

$$E\left[\max_{C,R} \int_0^\infty e^{-\rho t} U(C_t) dt\right] = \max_{C,R} \int_0^\infty \omega_T \left\{\int_0^T e^{-\rho t} U(C_t) dt + e^{-\rho t} W(K_T, Q_T)\right\} dT. \qquad (\text{E.6})$$

ω_t is the exogenous given probability that at time t the innovation date T has been reached:

$$prob(T = t) = \omega_t$$

$$\int_0^\infty \omega_t dt = 1$$

$$\omega_t > 0$$

$$\Omega_t = \int_t^\infty \omega_t dt.$$

Ω_t is the probability that at time t the innovation date T has not been reached. Partial integrating of the right side of equation E.6 yields:

$$E\left[\max_{C,R} \int_0^\infty e^{-\rho t} U(C_t) dt\right] = \max_{C,R} \int_0^\infty e^{-\rho t} [\Omega_t U(C_t) + \omega_t W(K_t, Q_t)] dt. \qquad (\text{E.7})$$

The starting inventory at the innovation date equals $Q_T = S_0 - \int_0^T R_t dt$, $Q_t = S_t$.

The expected present value of utility is maximized for optimal social welfare:

$$E\left[\max_{C,R}\int_0^\infty e^{-\rho t}U(C_t)dt\right] = \max_{C,R}\int_0^\infty e^{-\rho t}[\Omega_t U(C_t)+\omega_t W(K_t,S_t)]dt \quad \text{(E.8)}$$

subject to the dynamic capital accumulation, the process of resource extraction:

$$\dot{K} = F(K_t,R_t)-C_t$$
$$\dot{S} = -R_t$$

and the boundary conditions and non-negativity constraints:

$$K(0)=K_0 \, , \; K(t)\ge 0$$
$$S(0)=S_0 \, , \; S(t)\ge 0.$$

The present-value Hamiltonian in continuous time is defined as:

$$H = e^{-\rho t}[\Omega_t U(C_t)+\omega W(K_t,S_t)]+\psi_1[F(K_t,R_t)-C_t]+\psi_2[-R_t]. \quad \text{(E.9)}$$

Using subscripts to denote partial derivatives and omitting the argument t, the first-order conditions are given by:

$$H_C: e^{-\rho t}\Omega U'(C)-\psi_1 \overset{!}{=} 0 \qquad\qquad \text{(E.10)}$$
$$H_K: e^{-\rho t}\omega W_K+\psi_1 F_K \overset{!}{=} -\dot{\psi}_1 \qquad\qquad \text{(E.11)}$$
$$H_R: \quad \psi_1 F_R-\psi_2 \overset{!}{=} 0 \qquad\qquad \text{(E.12)}$$
$$H_S: \quad e^{-\rho t}\omega W_S \overset{!}{=} -\dot{\psi}_2. \qquad\qquad \text{(E.13)}$$

Differentiating E.10 with respect to time, dividing it by ψ_1, and combining the solution with E.11 yields the optimal consumption path (Ramsey's rule):

$$\dot{C} = -[F_K-\rho]\frac{U'(C)}{U''(C)} - \frac{\omega}{\Omega}\frac{U'(C)}{U''(C)}\left[\frac{W_K}{U'(C)}-1\right]. \qquad \text{(E.14)}$$

Note that the derivative of Ω with respect to time is negative $(-\omega)$. The probability that the innovation date has not been reached decreases with increasing time. ω_t/Ω_t is the conditional probability of the substitute being discovered at t given that it has not been discovered earlier.

For the case that households smooth consumption over time $(U(C) = ln(C))$ and the Cobb-Douglas production function $Y = F(K,R) = K^\alpha R^{1-\alpha}$, the optimal consumption path is given by:

$$\frac{\dot{C}}{C} = [F_K - \rho] + \frac{\omega}{\Omega}[W_K C - 1] \qquad (E.15)$$

$$\frac{\dot{C}}{C} = [\alpha x^{1-\alpha} - \rho] + \frac{\omega}{\Omega}[W_K C - 1]. \qquad (E.16)$$

Further, the Hotelling rule can be derived by differentiating E.12 with respect to time and combining this derivative with equation E.13 and E.11. This yields:

$$\frac{\dot{F}_R}{F_R} = \frac{\omega}{\Omega}\frac{W_K}{U'(C)} + F_K - \frac{\omega}{\Omega}\frac{W_S}{U'(C)F_R} \qquad (E.17)$$

$$\dot{x} = \frac{\omega}{\Omega}\frac{W_K}{U'(C)}\frac{1}{\alpha x^{-1}} + x^\alpha - \frac{\omega}{\Omega}\frac{W_S}{U'(C)}\frac{1}{\alpha(1-\alpha)x^{\alpha-1}}. \qquad (E.18)$$

If $W_S = W_K = 0$ the optimal paths simplify to:

$$\frac{\dot{C}}{C} = \alpha x^{1-\alpha} - \left(\rho + \frac{\omega}{\Omega}\right) \qquad (E.19)$$

$$\dot{x} = x^\alpha. \qquad (E.20)$$

The optimal capital to resource ratio is identical to that of the basic model.

For $\omega_t/\Omega_t = 0$ the optimal consumption path reduces to that of the Dasgupta-Heal model:

$$\dot{C} = [F_K - \rho]C - \frac{\omega}{\Omega}C. \qquad (E.21)$$

With the Poisson distribution $\omega_t = \pi e^{-\pi t}$, the optimal consumption path is given by:

$$\frac{\dot{C}}{C} = \alpha x^{1-\alpha} - \frac{\pi e^{-\pi t}}{e^{-\pi t}}$$

$$\frac{\dot{C}}{C} = \alpha x^{1-\alpha} - (\rho + \pi). \tag{E.22}$$

Note that the obtained optimal solutions only hold until the substitute is discovered. At the innovation date T the economy is totally "new". From then on it is optimal to pursue the optimal solution derived for the period beyond T.

E.1 A Certainty Equivalent Solution

The optimization problem for uncertain technological progress is stated differently:

$$\max_{C,R} \int_0^\infty e^{-\upsilon t} U(C_t) dt \tag{E.23}$$

subject to the dynamic capital accumulation, the process of resource extraction:

$$\dot{K} = F(K_t, R_t) - C_t$$
$$\dot{S} = -R_t$$

and the boundary conditions and non-negativity constraints:

$$K(0) = K_0 , \; K(t) \geq 0$$
$$S(0) = S_0 , \; S(t) \geq 0.$$

υ is a discount rate which is independent of the path followed by the economy.

Solving this optimization problem yields the basic Dasgupta-Heal results (A.14 and A.19), where $v \geq \rho$.

$$\dot{C} = [\alpha x_t^{\alpha-1} - v]C \qquad\qquad (E.24)$$

$$\dot{x} = x^{\alpha}. \qquad\qquad (E.25)$$

Then, necessary and sufficient conditions to obtain identical solutions for the problem under uncertainty and the certainty equivalent problem are:

1. $W_K = W_S = 0$ for all K,S
2. $v = \rho + \pi$.

ω is Poisson distributed, i.e. $\omega/\Omega = \pi$. Equation E.24 is equal to E.16 for $W_K = 0$ and $v = \rho + \pi$. In addition $E.25 = E.18$ is true when $W_K = 0$, then W_S is also zero:

$$\pi \frac{W_S}{U'(C)} \frac{1}{\alpha(1-\alpha)x^{\alpha-1}} + \pi \frac{W_K}{U'(C)} \frac{1}{\alpha x^{-1}} = 0. \qquad\qquad (E.26)$$

Under the previously stated conditions the optimal depletion problem can be solved by pretending that the new technology will never appear. Solely the utility discount rate is modified.

F   The Kamien-Schwartz Model

The optimization problem for the period until the technological breakthrough is reached can be stated as follows. Households will receive utility U(C) as long as the old technology is in use. The new technology becomes available during two points in time $(t, t + dt)$ with probability:

$$d\omega(z) = \omega'(z)\dot{z} = \omega'(z)v(m) ,\qquad\text{(F.1)}$$

providing a future utility stream with discounted value W at time t. Since technological progress is a process of innovation the probability depends on the cumulative effective effort z_t devoted to the project by time t. $\Omega(z_t) = 1 - \omega(z_t)$ is the probability that at time t the discovery date T of the new technology has not been reached.

The expected present value of utility is maximized for optimal social welfare:

$$E\left[\max_{C,R,m} \int_0^\infty e^{-\rho t} U(C_t)dt\right] = \max_{C,R,m} \int_0^\infty e^{-\rho t}[U(C_t)\{\Omega(z)\} + \omega'(z)v(m)W]dt \quad\text{(F.2)}$$

subject to the dynamic capital accumulation, the process of resource extraction, the process of innovation:

$$\dot{K} = F(K_t, R_t) - C_t - m_t$$
$$\dot{S} = -R_t$$
$$\dot{z} = v(m_t)$$

and the boundary conditions and non-negativity constraints:

$$K(0) = K_0 , \ K(t) \geq 0$$
$$S(0) = S_0 , \ S(t) \geq 0$$
$$z(0) = z_0 , \ z(t) \geq 0.$$

The present-value Hamiltonian in continuous time is defined as:

$$H = e^{-\rho t}[U(C_t)\Omega(z) + \omega'(z_t)v(m_t)W] + \psi_1[F(K_t,R_t) - C_t - m_t] + \psi_2[-R_t] + \psi_3 v(m_t) .$$
(F.3)

Using subscripts to denote partial derivatives and omitting the argument t, the first-order conditions are given by:

$$H_C : \qquad e^{-\rho t}U'(C)\Omega(z) - \psi_1 \qquad \overset{!}{=} 0 \qquad\qquad \text{(F.4)}$$

$$H_K : \qquad\qquad \psi_1 F_K \qquad\qquad \overset{!}{=} -\dot{\psi}_1 \qquad\qquad \text{(F.5)}$$

$$H_R : \qquad\qquad \psi_1 F_R - \psi_2 \qquad\quad \overset{!}{=} 0 \qquad\qquad \text{(F.6)}$$

$$H_S : \qquad\qquad\qquad 0 \qquad\qquad\quad \overset{!}{=} -\dot{\psi}_2 \qquad\qquad \text{(F.7)}$$

$$H_m : \quad e^{-\rho t}\omega'(z)v'(m)W - \psi_1 + \psi_3 v'(m) \ \overset{!}{\leq} 0 \quad for \ m \geq 0 \qquad \text{(F.8)}$$

$$H_z : \ -e^{-\rho t}U(C)\omega'(z) + e^{-\rho t}\omega''(z)v(m)W \overset{!}{=} -\dot{\psi}_3. \qquad \text{(F.9)}$$

Equation F.8 needs further discussion. There are two possible solutions for m_t which maximize the Hamiltonian at each point in time. One policy is that of not undertaking R&D, i.e. $m = 0$. Then at $t = 0$:

(i) $\psi_3(0)v'(0) - \psi_1(0) < 0,$

since $z = \omega'(z) = v(0) = 0.$

The other policy is to pursue R&D, i.e. $m > 0$. Then at $t = 0$:

(ii) $\psi_3(0)v'(m) - \psi_1(0) = 0,$

which is the case while a technological breakthrough is being researched.

To find the point where R&D begins, F.9 is integrated yielding:

$$\int_t^\infty \dot\psi_3 ds = \int_t^\infty e^{-\rho s}U(C)\omega'(z)ds - \int_t^\infty e^{-\rho s}\omega''(z)v(m)W ds$$

$$-\psi_3 = \int_t^\infty e^{-\rho s}U(C)\omega'(z)ds - \left[\omega'(z)We^{-\rho s}\right]_t^\infty - \int_t^\infty e^{-\rho s}\omega'(z)\rho W ds$$

$$\psi_3 = -e^{-\rho t}\omega'(z)W + \int_t^\infty e^{-\rho s}\omega'(z)[\rho W - U(C)]ds. \tag{F.10}$$

Using this result F.8 can be written as:

$$v'(m)\int_t^\infty e^{-\rho s}\omega'(z)[\rho W - U(C)]ds \leq \psi_1. \tag{F.11}$$

For $m > 0$ the marginal value of the composite good used in *R&D* must equal its marginal value in capital investment, equation F.11. Inserting this solution into (i) gives:

(*i*) $v'(0)\int_0^\infty e^{-\rho s}\omega'(z)[\rho W - U(C)]ds < \psi_1(0).$

This inequality holds true if *R&D* is never undertaken, since then $\omega'(z) = 0$ and thus $\psi_3(0) = 0$. However, if *R&D* is undertaken at some point in the future the expected net utility from innovation is positive. The marginal value of cumulative effective *R&D* effort is greater than zero ($\psi_3(0) > 0$). There must be a value of ψ_1 at a point t_0 where the equality holds true, i.e. (ii) holds true. To show this more easily (i) is rearranged to:

(*i'*) $\psi_3(0)v'(0) - \frac{\psi_2(0)}{F_R} < 0.$

F_R increases over time, i.e. the value of the fraction decreases. There will be a point t_0 where (i') becomes zero. At this point the marginal value of the composite good used in *R&D* equals the marginal value of the composite good in capital investment. The innovation process begins. For the case before the onset of R&D the result from chapter 3 can be applied. The case of undertaking R&D is analyzed in the following.

Equality (ii) must be true during *R&D*. For the case that households smooth consumption over time ($U(C) = ln(C)$) and using the Cobb-Douglas production functions $Y = F(K,R) = K^\alpha R^{1-\alpha}$, the optimal path of the capital to resource ratio is obtained by differentiation F.6 with respect to time and combining the solution with F.7 and F.5:

$$\dot{x} = x^\alpha. \tag{F.12}$$

This solution is identical to the basic Dasgupta-Heal result. In addition, differentiating F.4 with respect to time, dividing it by ψ_1 and combining the solution with F.5 yields the optimal consumption path (Ramsey's rule):

$$\frac{\dot{\psi}_1}{\psi_1} = -\rho + \frac{U''(C)\dot{C}}{U'(C)} - \frac{\omega'(z)\dot{z}}{1-\omega(z)}$$

$$-F_K = -\rho + \frac{U''(C)\dot{C}}{U'(C)} - h(z)v(m). \tag{F.13}$$

This simplifies in the case of a logarithmic utility function to:

$$\frac{\dot{C}}{C} = \alpha x^{\alpha-1} - \rho - v(m)h(z). \tag{F.14}$$

To obtain the optimal effort in R&D F.8 is differentiated with respect to time yielding:

$$\dot{\psi}_1 = \left[-\rho e^{-\rho t}\omega'(z)W + e^{-\rho t}\omega''(z)\dot{z}W + \dot{\psi}_3\right]v'(m) + \left[e^{-\rho t}\omega'(z)W + \psi_3\right]v''(m)\dot{m}. \tag{F.15}$$

Using this result with F.9 and inserting ψ_3 gives:

$$\dot{\psi}_1 = \left[-\rho e^{-\rho t}\omega'(z)W + e^{-\rho t}U(C)\omega'(z)\right]v'(m) + \left[e^{-\rho t}\omega'(z)W + \psi_3\right]v''(m)\dot{m}$$

$$\dot{\psi}_1 = \left[e^{-\rho t}\omega'(z)[U(C) - \rho W]\right]v'(m) + \psi_1 \frac{v''(m)\dot{m}}{v'(m)}. \tag{F.16}$$

This equation with F.4 and F.5 yields the optimal effort in R&D, given by:

$$-\frac{v''(m)\dot{m}}{v'(m)} = \alpha x^{\alpha-1} - v'(m)h(z)C[\rho W - ln(C)], \qquad (F.17)$$

where h(z) denotes the conditional probability of completion.[35]

Note that the obtained optimal solutions are only true until the substitute is discovered. At the innovation date T there is a switch. From then on it is optimal to pursue the optimal solution derived in the Dasgupta-Heal model for the period beyond T.[36]

F.1 Phase Diagram

The behavior of C and m are analyzed separately at date t_m. The behavior of $\dot{C}$ is described by:

$$\frac{\dot{C}}{C} = F_K - \rho - v(m)h(z). \qquad (F.18)$$

$v(m)h(z)$ is either positive or zero. $\dot{C}$ increases or decreases:

a) if $F_K \leq \rho$ then $\dot{C} \leq 0$ for $h(z) \geq 0$
b) if $F_K > \rho$ then $\dot{C} \lessgtr 0$ as $m(t) \gtrless m^0(t)$ for $h(z) > 0$.

---

[35] Compare Appendix F.2 for a formal derivation of the conditional probability of completion.
[36] Compare Appendix E.

$m^0(t)$ is implicitly defined by equation F.18 with $F_K = \alpha x^{\alpha-1}$:

$$v(m^0(t)) = \frac{\alpha x^{\alpha-1} - \rho}{h(z)}, \tag{F.19}$$

specifying the $\dot{C}$ locus in the C-m space. The movement of this locus over time is given by:

$$v'(m)dm^0/dt = -\frac{\alpha(1-\alpha)x^{\alpha-2}\dot{x}}{h(z)} - \frac{(\alpha x^{\alpha-1} - \rho)h'(z)\dot{z}}{h(z)^2}. \tag{F.20}$$

Since $v'(m)$ is positive, $m^0(t)$ decreases over time. The $\dot{C}$ locus moves down.

The behavior of $\dot{m}$ is described by:

$$-\frac{v''(m)\dot{m}}{v'(m)} = F_K - v'(m)h(z)C[\rho W - ln(C)]. \tag{F.21}$$

$\dot{m}$ increases or decreases:

a) if $h(z) = 0$ then $\dot{m} \geq 0$
b) if $\frac{\alpha x^{\alpha-1}}{v'(0)h(z)} > G(C^*)$ then $\dot{m} > 0$ for $h(z) > 0$
c) if $\frac{\alpha x^{\alpha-1}}{v'(0)h(z)} < G(C^*)$ then $\dot{m} \gtreqqless 0$ as $m(t) \gtreqqless m^C(C,t)$ for $h(z) > 0$.

Case b) and c) need further discussion. For $h(z) > 0$ and $\dot{m} = 0$ the following must hold true:

$$\frac{\alpha x^{\alpha-1}}{v'(m)h(z)} = C[\rho W - ln(C)]. \tag{F.22}$$

The left side of equation F.22 increases with m $\left(v'(m) = \frac{1}{2m^{0.5}}\right)$, while the right hand side is a concave function of C, i.e. G(C). The smallest value of the left side at any time is given by $\frac{x^{\alpha-1}}{v'(0)h(z)}$. G(C) is maximized at $C^* = e^{\rho W-1}$. As long as $\frac{x^{\alpha-1}}{v'(0)h(z)} > C^*$ the effort in R&D increases (case b). If this relationship is not true, then the effort in R&D behaves according to the shape of the $\dot{m} = 0$ locus (case c).

Equation F.22 is considered an implicit definition of the function $m^C(C,t) = m$. Holding t fixed and differentiating F.22 with respect to C yields:

$$-\frac{\alpha x^{\alpha-1} v''(m)}{(v'(m))^2 h(z)} \frac{\partial m^C}{\partial C} = G'(C). \tag{F.23}$$

Since $v''(m)$ is assumed to be negative, $\partial m^C / \partial C$ takes the sign of G'(C). The following relationship holds true:

$$\frac{\partial m^C}{\partial C} > 0 \quad for\ 0 < C < C^*$$
$$\frac{\partial m^C}{\partial C} < 0 \quad for\ C^* < C.$$

Considering equation F.21 with this previously stated result, the following must hold true. In the C-m space above the $\dot{m} = 0$ locus $\dot{m} > 0$ and $\dot{m} < 0$ for points below the locus.

The movement of the $\dot{m} = 0$ locus over time can be obtained by differentiating F.22 with respect to time (holding C constant). This yields:

$$\frac{-\alpha(1-\alpha)x^{\alpha-2}\dot{x}v'(m)h(z)}{[v'(m)]^2[h(z)]^2} - \frac{\alpha x^{\alpha-1}v'(m)h'(z)\dot{z}}{[v'(m)]^2[h(z)]^2} - \frac{\alpha x^{\alpha-1}v''(m)h(z)\frac{\partial m^C}{\partial t}}{[v'(m)]^2[h(z)]^2} = 0$$

$$-\frac{\alpha x^{\alpha-1}v''(m)\frac{\partial m^C}{\partial t}}{[v'(m)]^2 h(z)} = \frac{\alpha(1-\alpha)x^{\alpha-2}\dot{x}}{v'(m)h(z)} + \frac{\alpha x^{\alpha-1}h'(z)\dot{z}}{v'(m)[h(z)]^2}. \tag{F.24}$$

The left side is positive since $v''(m) < 0$. The m-coordinate of each point of the $\dot{m}$ locus increases over time. The $\dot{m}$ locus moves up. The C-coordinate of its peak remains fixed at C^*. The intercepts of the curve on the $m = 0$ axis satisfy:

$$\frac{\alpha x^{\alpha-1}}{v'(0)h(z)} = G(C). \tag{F.25}$$

Since the left side of equation F.25 decreases through time, the smaller interception point with the $m = 0$ axis, $\underline{C}$, decreases over time while the larger one, $\bar{C}$, increases.

F.2 Conditional Probability of Completion

The conditional probability of completion can be derived as follows. Consider the probability, that R&D will be completed in the small time interval $t + dt$ with incremental effort z. $\omega(z_t)$ is the probability that the R&D will be successfully completed by the time cumulative effort is z ($P\{z_T \leq z_t\}$). $\Omega(z_t) = 1 - \omega(z_t)$ is the probability that at time t the discovery date T of the new technology has not been reached ($P\{z_t < z_T\}$). Using the multiplication rule of probability the following holds true:

$$P\{z_t \leq z_T < z_{t+dt}\} = P\{z_T \geq z_t\}P\{z_T < z_{t+dt} \mid z_T \geq z_t\}. \tag{F.26}$$

Further if $\Omega(z_t) = P\{z_t < z_T\}$ is positive:

$$P\{z_T < z_{t+dt} \mid z_T > z_t\} = \frac{P\{z_t < z_T < z_{t+dt}\}}{P\{z_T > z_t\}}$$

$$P\{z_T < z_{t+dt} \mid z_T > z_t\} = \frac{P\{z_t < z_T < z_{t+dt}\}}{\Omega(z_t)}. \tag{F.27}$$

Now the conditional rate of completion for the interval $(t, t + dt)$ is the conditional probability of completion in the interval (given that the technological breakthrough has not been reached at t) divided by the length of the interval. Thus, the conditional probability of completion is given by:

$$h(z_t) = \frac{P\{z_T < z_{t+dt} \mid z_T > z_t\}}{z_{t+dt} - z_t} = \frac{P\{z_t < z_T < z_{t+dt}\}}{\Omega(z_t)(z_{t+dt} - z_t)}$$

$$h(z_t) = \frac{\Omega(z_t) - \Omega(z_{t+dt})}{\Omega(z_t)(z_{t+dt} - z_t)}$$

$$h(z_t) = \frac{\Omega(z_t) - \Omega(z_{t+dt})}{\Omega(z_t)(dz)}. \tag{F.28}$$

For very little intervals ($\lim_{dz \to 0}$), $h(z_t)$ is approximately given by:

$$h(z_t) = \frac{-\Omega'(z_t)}{\Omega(z_t)}$$

$$h(z_t) = \frac{\omega'(z_t)}{\Omega(z_t)}. \tag{F.29}$$

References

Acemoglu, D. (2002). Directed Technical Change. *The Review of Economic Studies*, 69(4):781–809.

Acemoglu, D. and Aghion, P. (2012). The Environment and Directed Technical Change. *The American Economic Review*, 102(1):131–166.

Ayres, R. U. (1999). The Second Law, the Fourth Law, Recycling and Limits to Growth. *Ecological Economics*, 29(3):473–483.

Ayres, R. U. and Kneese, A. V. (1969). Production, Consumption, and Externalities. *The American Economic Review*, pages 282–297.

Barbier, E. B. (1999). Endogenous Growth and Natural Resource Scarcity. *Environmental and Resource Economics*, 14(1):51–74.

Barro, R. and Sala-i Martin, X. (2004). *Economic Growth*. McGraw-Hill, 2nd edition.

BP (2014). BP Statistcial Review of World Energy. http://www.bp.com/en/global/corporate/about-bp/energy-economics/statistical-review-of-world-energy.html.

Bretschger, L. (1998). How to Substitute in Order to Sustain: Knowledge Driven Growth Under Environmental Restrictions. *Environment and Development Economics*, 3(04):425–442.

Bretschger, L. and Smulders, S. (2004). Sustainability and Substitution of Exhaustible Natural Resources. How Resource Prices Affect Long-term R&D-Investments. Technical report, CER-ETH-Center of Economic Research at ETH Zurich.

Cass, D. (1965). Optimum Growth in an Aggregative Model of Capital Accumulation. *The Review of Economic Studies*, pages 233–240.

Cobb, C. W. and Douglas, P. H. (1928). A Theory of Production. *The American Economic Review*, pages 139–165.

Cohen, J. E. (1995). Human Carrying Capacity. *Science*, 269:341.

Dasgupta, P. and Heal, G. (1974). The Optimal Depletion of Exhaustible Resources. *The Review of Economic Studies*, pages 3–28.

Deutsche Bank (2013). A Guide to the Oil & Gas Industry. http://www.wallstreetoasis.com/files/DEUTSCHEBANK-AGUIDETOTHEOIL%EF%BC%86GASINDUSTRY-130125.pdf.

Di Vita, G. (2001). Technological Change, Growth and Waste Recycling. *Energy Economics*, 23(5):549–567.

Di Vita, G. (2005). Renewable Resources and Waste Recycling. *Environmental Modeling & Assessment*, 9(3):159–167.

Di Vita, G. (2006). Natural Resources Dynamics: Exhaustible and Renewable Resources, and the Rate of Technical Substitution. *Resources policy*, 31(3):172–182.

Di Vita, G. (2007). Exhaustible Resources and Secondary Materials: A Macroeconomic Analysis. *Ecological Economics*, 63(1):138–148.

Gaudet, G. (2007). Natural Resource Economics Under the Rule of Hotelling. *Canadian Journal of Economics*, 40(4):1033–1059.

Grimaud, A. and Rougé, L. (2003). Non-Renewable Resources and Growth with Vertical Innovations: Optimum, Equilibrium and Economic Policies. *Journal of Environmental Economics and Management*, 45(2):433–453.

Grimaud, A. and Rougé, L. (2005). Polluting Non-Renewable Resources, Innovation and Growth: Welfare and Environmental policy. *Resource and Energy Economics*, 27(2):109–129.

Grossman, G. and Helpman, E. (1991). *Innovation and Growth in the Global Economy*. Cambridge.

Groth, C. (2006). A New-Growth Perspective on Non-Renewable Resources. Discussion Papers 06-26, University of Copenhagen. Department of Economics.

Hartwick, J. M. (1977). Intergenerational Equity and the Investing of Rents from Exhaustible Resources. *The American Economic Review*, pages 972–974.

Hartwick, J. M., Van Long, N., and Tian, H. (2003). On the Peaking of Consumption with Exhaustible Resources and Zero Net Investment. *Environmental and Resource Economics*, 24(3):235–244.

Havranek, T. (2014). Measuring Intertemporal Substitution: The Importance of Method Choices and Selective Reporting. *Czech National Bank and Charles University, Prague.*

Heijman, W. J. M. (1991). *Depletable Resources and the Economy.* Agricultural University.

Hotelling, H. (1931). The Economics of Exhaustible Resources. *The Journal of Political Economy*, pages 137–175.

Inada, K.-I. (1963). On a Two-sector Model of Economic Growth: Comments and a Generalization. *The Review of Economic Studies*, pages 119–127.

Just, R. E., Netanyahu, S., and Olson, L. J. (2005). Depletion of Natural Resources, Technological Uncertainty, and the Adoption of Technological Substitutes. *Resource and Energy Economics*, 27(2):91–108.

Kamien, M. I. and Schwartz, N. L. (1978). Optimal Exhaustible Resource Depletion with Endogenous Technical Change. *The Review of Economic Studies*, pages 179–196.

Kemp, M. C. and Long, N. (1980). *Exhaustible Resources, Optimality, and Trade: The Firm As Resource-Farmer.* Amsterdam: North-Holland.

Kneese, A. V., Ayres, R. U., d'Arge, R. C., et al. (1970). *Economics and the Environment: A Materials Balance Approach.* London: the Johns Hopkins Press.

Kolstad, C. D. and Krautkraemer, J. A. (1993). Natural Resource Use and the Environment. *Handbook of Natural Resource and Energy Economics*, 3:1219–1265.

Krautkraemer, J. A. (1998). Nonrenewable Resource Scarcity. *Journal of Economic Literature*, pages 2065–2107.

Kuhn, T., Pittel, K., and Schulz, T. (2003). Recycling for Sustainability-A Long Run Perspective? *International journal of global environmental issues*, 3(3):339–355.

Laibson, D. (1997). Golden Eggs and Hyperbolic Discounting. *The Quarterly Journal of Economics*, pages 443–477.

Lucas, R. (1988). On the Mechanics of Economic Development. *Journal of Monetary Economics*, 22:3–42.

Mäler, K.-G. E. E. (1974). A Theoretical Inquiry. *Baltimore: John Hopkins University*.

Malthus, T. R. (1803). An Essay on the Principle of Population or a View of its Past and Present Effects on Human Happiness. Technical report, Revised and Expanded 2nd Edition of 1798.

Meadows, D. H., Meadows, D. L., Randers, J., and Behrens, W. W. (1972). The Limits to Growth. *New York*, 102.

Miller, E. (2008). *An Assessment of CES and Cobb-Douglas Production Functions*. Congressional Budget Office.

National Oceanic and Atmospheric Administration (2011). An Oil Spill Primer for Students. http://oceanservice.noaa.gov/education/stories/oilymess/supp_primer.html.

Nordhaus, W. D., Houthakker, H., and Solow, R. (1973). The Allocation of Energy Resources. *Brookings Papers on Economic Activity*, pages 529–576.

Perman, R., Yue, M., McGilvray, J., and Common, M. (2003). *Natural Resource and Environmental Economics*. Pearson Education, 3rd edition.

Pezzey, J. and Withagen, C. A. (1998). The Rise, Fall and Sustainability of Capital-Resource Economies. *The Scandinavian Journal of Economics*, 100(2):513–527.

Pittel, K., Amigues, J.-P., and Kuhn, T. (2006). Long-Run Growth and Recycling: A Material Balance Approach. Technical report, CER-ETH-Center of Economic Research at ETH Zurich.

Pittel, K., Amigues, J.-P., and Kuhn, T. (2010). Recycling Under a Material Balance Constraint. *Resource and Energy Economics*, 32(3):379–394.

PlasticsEurope (2014). Plastics - Let's Talk About It. http://www.plasticseurope.de/informationszentrum/publikationen.aspx.

PlasticsEurope (2015). Plastics - the Facts 2014/2015. `http://www.plasticseurope.de/informationszentrum/publikationen.aspx`.

Ramsey, F. P. (1928). A Mathematical Theory of Saving. *The Economic Journal*, pages 543–559.

Rawls, J. (1971). *A Theory of Justice*. Oxford University Press, Cambridge.

Rebelo, S. (1991). Long-Run Policy Analysis and Long-Run Growth. *The Journal of Political Economy*, 99(3):500–521.

Romer, P. M. (1990). Endogenous Technological Change. *Journal of Political Economy*, 98(5 pt 2).

Schou, P. (2000). Polluting Non-Renewable Resources and Growth. *Environmental and Resource Economics*, 16(2):211–227.

Schulze, W. D. (1974). The Optimal Use of Non-Renewable Resources: The Theory of Extraction. *Journal of Environmental Economics and Management*, 1(1):53–73.

Smulders, S. and Withagen, C. (2012). Green Growth-Lessons From Growth Theory. *World Bank Policy Research Working Paper*, 6230.

Solow, R. M. (1974). Intergenerational Equity and Exhaustible Resources. *The Review of Economic Studies*, pages 29–45.

Stiglitz, J. (1974). Growth with Exhaustible Natural Resources: Efficient and Optimal Growth Paths. *The Review of Economic Studies*, pages 123–137.

Tahvonen, O. and Salo, S. (2001). Economic Growth and Transitions Between Renewable and Nonrenewable Energy Resources. *European Economic Review*, 45(8):1379–1398.

Tietenberg, T. H. and Lewis, L. (2012). *Environmental & Natural Resource Economics*. Pearson Addison Wesley, 9th edition.

Tsur, Y. and Zemel, A. (2003). Optimal Transition to Backstop Substitutes for Nonrenewable Resources. *Journal of Economic Dynamics and Control*, 27(4):551–572.

Tsur, Y. and Zemel, A. (2005). Scarcity, Growth and R&D. *Journal of Environmental Economics and Management*, 49(3):484–499.

Wacker, H. (1987). *Rezyklierung als Intertemporales Allokationsproblem in Gesamtwirtschaftlichen Planungsmodellen*. Frankfurt am Main: Lang.

Weil, D. (2013). *Economic Growth*. Pearson Addison Wesley, 3rd edition.

Weinstein, M. C. and Zeckhauser, R. J. (1974). Use Patterns for Depletable and Recycleable Resources. *The Review of Economic Studies*, pages 67–88.

World Resource Institute (2010). Agriculture's Share of Global Environment Impact (2010). http://www.wri.org/sites/default/files/uploads/ag_environmental_impact_0.jpg.